Introduction to Diffusion Tensor Imaging

Introduction to Diffusion Tensor Imaging

by

Susumu Mori
Johns Hopkins University School of Medicine
Department of Radiology
Division of MRI Research and
F.M. Kirby Research Center for Functional Brain Imaging
Kennedy Krieger Institute
Baltimore, MD
USA

ELSEVIER
Amsterdam ● Boston ● Heidelberg ● London ● New York ● Oxford
Paris ● San Diego ● San Francisco ● Singapore ● Sydney ● Tokyo

Elsevier
Radarweg 29, PO Box 211, 1000 AE Amsterdam, The Netherlands
Linacre House, Jordan Hill, Oxford OX2 8DP, UK

First edition 2007

British Library Cataloguing in Publication Data
A catalogue record for this book is available from the British Library

Library of Congress Cataloging in Publication Data
A catalog record for this book is available from the Library of Congress

ISBN: 978-0-444-52828-5

For information on all book publications
visit our website at books.elsevier.com

Printed and bound in the United Kingdom

Transferred to Digital Print 2009

Working together to grow
libraries in developing countries

www.elsevier.com | www.bookaid.com | www.sabre.com

ELSEVIER   BOOK AID
             International   Sabre Foundation

Contents

Preface

The root of diffusion tensor imaging dates back to 1965, when Drs. Stejskal and Tanner measured the diffusion constant of water molecules using nuclear magnetic resonance (NMR) and magnetic field gradient systems. Compared to diffusion measurement using chemical tracers, the NMR-based method has several unique aspects. First, it is noninvasive. Second, it measures molecular motion along an arbitrary, predetermined axis; we can measure water diffusion along right–left, fore–aft, up–down, or any oblique angle we wish. If we are measuring freely diffusing water, this unique capability does not mean much, because measurements along any orientation give the same result. This is what we call **isotropic** diffusion. However, the situation changes when we study biological tissues such as muscle and brain, which consist of fibers with coherent orientations. In such systems, water tends to diffuse along the fiber, and diffusion becomes **anisotropic**. This means that the results of diffusion measurements are not the same if they are measured along different orientations. When diffusion is measured along, for example, muscle fiber, the diffusion constant becomes largest, it becomes smallest when measured perpendicular to the fiber.

We are interested in diffusion anisotropy because we can deduce the anatomy of the sample from it. For muscle tissue, we usually have *a priori* knowledge about the fiber anatomy, such that we can measure diffusion along or perpendicular to the fiber. However, what if we do not know the fiber orientation? In theory, we should be able to determine the fiber orientation by measuring diffusion constants along many orientations. The fiber should align to the measurement orientation with the largest diffusion constant. One such clear example is the measurement of the brain's diffusion anisotropy. We know that the brain contains axonal fibers, but their structures are highly complicated, and we sometimes do not fully understand them. It would be of great benefit, therefore, if we could measure diffusion anisotropy in the brain, from which we could deduce fiber structures. This is, however, a difficult challenge. Unlike a piece of muscle tissue, which consists of fibers of uniform orientation, different regions of the brain have different fiber orientations. Unless we cut out a small piece of the brain, the brain structure is too complicated to be described by simple diffusion constant measurements. We cannot, of course, remove pieces of brain tissues from a living human at will, but we can solve this problem by combining the diffusion measurements with magnetic resonance imaging (MRI), from which we can measure diffusion constants at each pixel with a resolution of a few millimeters. This technique is called **diffusion MRI**. In the late 1980s, application of this technique was started in brain studies using human and animal models. Soon, evidence of diffusion anisotropy in the brain began to surface.

Diffusion MRI provided a way of estimating brain fiber structures, using water diffusion properties as a probe. However, there was still a hurdle to overcome: how could we obtain useful quantitative parameters to describe diffusion anisotropy and fiber orientations from a series of diffusion measurements? If we have infinite scanning time, we can measure diffusion along thousands of orientations, from which we can identify the axis with the largest diffusion constant. However, scanning times are often limited and so is the number of measurement orientations. We needed a mathematical model to take this limited number of measurements and quantify such properties as fiber angles and the extent of anisotropy. In the early 1990s, various attempts were made, and a model based on tensor calculation emerged as one of the most widely adopted. The diffusion MRI based on this model is called **diffusion tensor imaging** or DTI (for one of the first papers, please see references #2 and #3 of this section).

Our cerebral hemispheres consist of approximately 100 billion neurons and 10 to 50 times more astrocytes. Almost all the neurons reside in the gray matter, which is rich in vasculatures and has complicated architectures with various types of neuronal cell bodies, dendrites, and axons. Neurons can communicate with other neurons via their dendrites for local communication and via axons that extend much further than the dendrites. One neuron has only one axon, but the axon may branch to communicate with multiple regions. Axons with similar destinations often form a huge bundle, called white matter tracts. Some of the prominent bundles such as the corpus callosum can easily be seen in postmortem brain dissections or conventional MRI. The brain white matter consists of these axonal bundles. These bundles are named according to their destinations. For example, those connecting two cortical regions are called U-fiber (between adjacent gyri), association fibers (between different lobes), or commissural fibers (between right and left hemispheres). Those connecting cortex and deep-brain regions (e.g., the cortex and thalamus or cortex and spinal cord) are called projection fibers.

When we perform DTI, image resolution is typically about 2 mm. If the observation window is of this size, it is understandable that the fiber structure of the gray matter looks incoherent and water diffusion looks isotropic because of its structural complexity. On the other hand, many regions in the white matter have axonal bundles larger than 2 mm. In these regions, water diffusion is anisotropic, with one preferential axis to diffuse. It is important to understand that low-diffusion anisotropy does not mean lack of fibers (the gray matter is full of fibers). It means lack of coherent fiber organization within the pixel size.

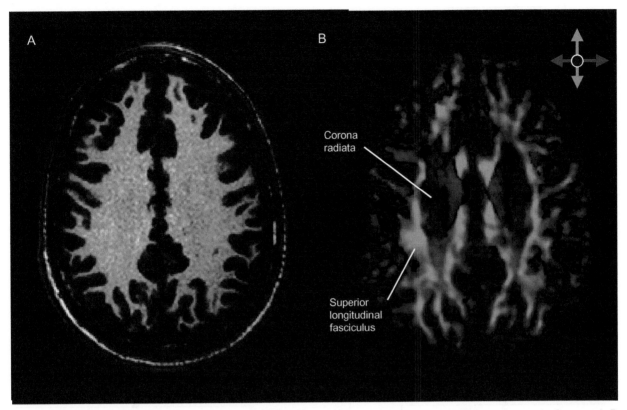

Fig. 1 Comparison between a conventional MRI (T_1-weighted image) and a DTI-based map (color map). In the color map, color represents fiber orientations; red, green, and blue represent fibers running along the right–left, anterior–posterior, and superior–inferior orientations.

There are two reasons why DTI is an important imaging modality. First, conventional MRI cannot reveal detailed anatomy of the white matter. Conventional MRI based on relaxation time relies on differences in chemical composition for their contrasts. For T_1- and T_2-weighted images, the amount of myelin plays a major role in differentiating the gray and white matter. However, the white matter looks quite homogeneous because it is homogeneous in terms of the chemical composition. In contrast, DTI can generate contrasts that are sensitive to fiber orientations. As an example, a color-coded fiber orientation map is shown in Fig. 1B. This image carries rich information about intra-white-matter axonal anatomy, which cannot be seen in the T_1-weighted image (Fig. 1A). By comparing with an existing anatomic atlas, we can identify where, for example, the so called "corona radiata" and "superior longitudinal fasciculus" are. The second reason is that, even after decades of anatomical studies of the human brain, our understanding of its connectivity is far from complete. There are many pathological conditions in which abnormalities in specific connections are suspected, but are difficult to delineate. It is anticipated that DTI may provide new information about human brain connectivity.

One of the purposes of this book is to explain how DTI works. To this end, it is important to know that there are two steps in

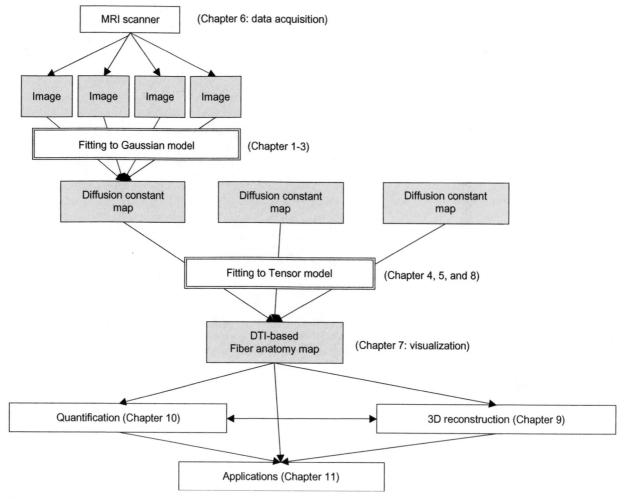

Fig. 2 Structure of this book with its relationship to the flow of diffusion tensor imaging.

DTI where we rely on **models** (i.e., in which we make assumptions and simplifications) (Fig. 2). The first step is when we calculate diffusion constants. Since the only information we can obtain from MRI scanners is grayscale images and the intensity at each pixel is determined by many factors such as proton density (water concentration), T_1 and T_2 relaxation, and water diffusion properties, the pixel intensity does not tell us what the diffusion constant is at each pixel. Chapter 1 to Chapter 3 explain how we can convert pixel intensities to a diffusion constant based on the **Gaussian diffusion model**. In the second step, we extract fiber anatomy information from a set of diffusion constant measurements based on the **tensor model**. This step is explained in Chapter 4 and Chapter 5. In both steps, we need multiple MR images with different experimental conditions and fit the raw intensity information to the models. It is very important to understand how the mere grayscale information from MR scanners is converted to fiber anatomy.

Other chapters are dedicated to data acquisition (Chapter 6), visualization (Chapter 7), non-tensor approaches (Chapter 8), 3D tract reconstruction (Chapter 9), quantification (Chapter 10), and applications (Chapter 11) as shown in Fig. 2.

ACKNOWLEDGEMENTS

Some sections of this textbook are based on two previous review articles; "Fiber tracking: Principles and strategies – a technical review", *NMR Biomed.* 15, 468–480 (2002) and "Principles of diffusion tensor imaging and its applications to basic neuroscience research", *Neuron*, 51, 527–540 (2006). Some figures are reproduced from these articles with permission. I would like to thank my coauthors of these articles, Drs. Peter C.M. van Zijl and Jiangyang Zhang for kindly allowing to use the materials. I also appreciate help from Dr. Hao Huang, Dr. Hangyi Jiang, Mr. Kegang Hua, Mr. Can Ceritoglu, and Mr. Jonathan A.D. Farrell to prepare texts and figures. I would like to thank my friends and colleagues who critically read the manuscript and gave me numerous valuable suggestions. They are Drs. Andrew L. Alexander, Hiroshi Ashikaga, Peter B. Barker, Christian Beaulieu, Fernando Calamante, and Allen W. Song. Some figures in this book were provided courtesy of my colleagues, Drs. Michael I. Miller, Peter A. Calabresi, Daniel S. Reich, Alexander H. Hoon, Elaine Stashinko, Daniel F. Hanley, and Andreas R. Luft. I would also like to thank Human Brain Project, National Institute of Aging, National Institute of Biomedical Imaging and BioEngineering, and National Center for Research Resources, NIH for financial support.

Chapter 1

Basics of diffusion measurement

1.1   NMR SPECTROSCOPY AND MRI CAN DETECT SIGNALS FROM WATER MOLECULES

Nuclear magnetic resonance (NMR) spectroscopy and magnetic resonance imaging (MRI) can be used to observe signals from various nuclei, but this book focuses only on the proton (^{1}H) of water molecules (H_2O) as the nucleus of interest. This is because more than 90% of protons in the body are located in water molecules, and the MRI signal is therefore dominated by water.

When we perform NMR spectroscopy, we put our sample (in this case, water) in a tube and place it inside a magnet. We then input energy into the sample (a process called excitation) and observe the signal emitting from the sample. This is similar to ringing a bell and listening to the sound. There are only three types of information in the signal: frequency (i.e., high or low pitch), intensity (i.e., loudness), and phase. To visualize this information, we use a waveform diagram such as that shown in Fig. 1.1. These waves are often called time-domain data because the horizontal axis is the time elapsed while we are listening to the signal. After Fourier transform, the time axis transforms to the so-called frequency domain, in which the horizontal axis is frequency. Figure. 1.1 shows a simple relationship between the time domain wave properties and frequency domain peak properties.

These three types of wave properties carry much information about the material; in our example of the bell, we can tell the size of the bell and its wall thickness. The signal that comes from the NMR sample after excitation is not sound but magnetic field oscillation. However, the situation is the same in that the only information we get is the three properties of oscillating signal, and they carry information about the sample properties. For example, if we put ethanol in the tube, we get a different frequency from that of water. From the frequency, we can tell much about what is in the tube. The intensity is proportional to the concentration of the sample. Unless we carry out advanced NMR spectroscopy or flow/velocity measurements, we do not use phase information. For us, the situation is very simple. We know that the sample is water, so we expect only one frequency which corresponds to that of water molecules. The concentration of water is 100%, so we should get a very intense signal.

The situation is also simple for MRI (Fig. 1.2). Here again, we have only intensity information. Because we are observing the

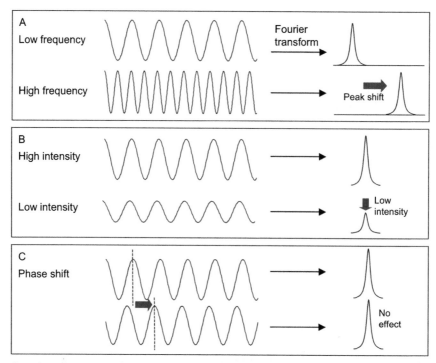

Fig. 1.1 Diagram showing frequency, intensity, and phase of waves. After Fourier transform, the waves in the time axis are transformed into peaks in the frequency axis. When we use so-called magnitude calculation, phase difference makes no impact on the peaks in the frequency domain.

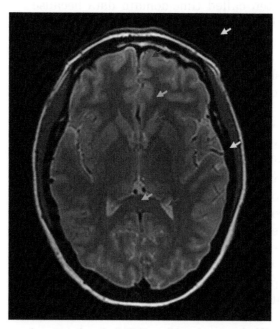

Fig. 1.2 An example of MR image (proton density). Yellow, blue, and red arrows represent regions with no, low, and high water concentrations, respectively.

water signal, we can assume that we have only one frequency. Phase information is usually discarded in MRI. When there is no water within a pixel, that pixel has no intensity. When there is a small amount of water like skin, it looks faint. The brain is full of water and thus looks very intense. Basically, it tells us where water is and what its concentration is (called proton density).

The question now is: how can we measure the diffusion constant of water molecules using NMR spectroscopy or MRI? We must somehow relate signal intensity information to the water diffusion process. Obviously, we cannot measure it from just one spectrum or image because, as mentioned earlier, the only information we have is signal intensity, which is mostly related to proton density (but not to the property of water diffusion). Before we begin a discussion of how to measure diffusion, it is important to first consider what "diffusion" is.

1.2   WHAT IS DIFFUSION?

We are interested in the motion of water molecules, and there are several ways in which they can move. For example, when we are imaging a human subject, he or she may move during the scan. This is the bulk movement of all water molecules in the body (Fig. 1.3A). Let us define bulk motion as the movement of water molecules that are more than a pixel in dimension. If our image resolution is 2 mm and a subject (and the water inside) moves more than 2 mm, this is the bulk motion. There are two types of effects on images: if the motion occurs during imaging, the image would be corrupted (blurring and ghosting). If we are acquiring multiple images, images before and after the motion would not be coregistered. We can therefore detect bulk motion by MRI, but, of course, this is not what we are interested in.

The second kind of water motion is flow (Fig. 1.3B), which is a one-directional water motion. If the amount of movement is larger than the pixel size (e.g., 2 mm), it can be treated as bulk motion. The term "flow" in this book specifically means one-directional water motion within a pixel. If we drop ink within a pixel (assuming that we can do such a thing), the ink stays in the pixel and the center of the ink moves. This flow has a distinctive effect on an MR signal. As will be explained later, it shifts the phase of the signal. Therefore, in principle, we can measure flow by NMR/MRI. However, as mentioned earlier, we usually discard phase information. So we can say that small flow motion within a pixel can be ignored. In practical situations, things can be more complicated because of large blood flow or non-unidirectional flow due to convoluted capillary structure. We usually ignore this because the population of water within blood vessels of the brain is small (about 5%) compared to those in the parenchyma. Also, when there is a tiny amount of bulk motion such as brain pulsation, it shifts the locations of water molecules on a subpixel scale. This also causes significant signal phase shift, which may interfere with our MRI measurements. This point will be described in more depth in Chapter 6.

The third kind of motion, and the one we are interested in, is "diffusion" (Fig. 1.3C). This is also called intra-voxel incoherent motion (IVIM), random motion, or Brownian motion. This

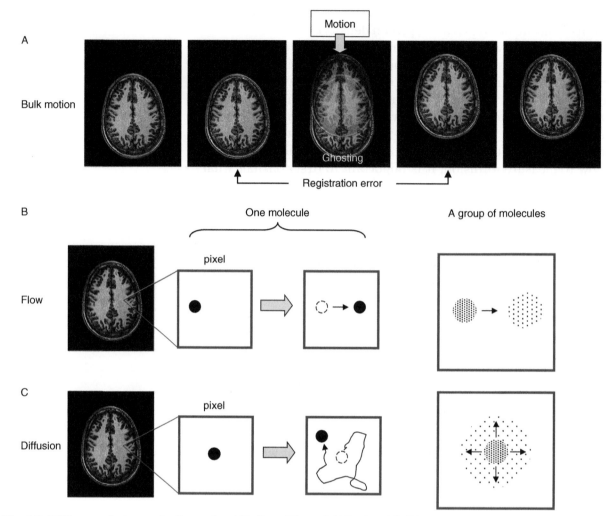

Fig. 1.3 Difference between bulk motion (A), flow (B), and diffusion (C). The actual amount of water diffusion is approximately 5–10 μm during MR measurements, while the pixel size is typically 2–5 mm. So, the water motion in (C) is exaggerated.

motion has nothing to do with physiological motion; even a water sample in a tube can move around unless it is frozen. If we drop ink in such a system, its shape become bigger as time elapses but its center remains at the same position. Along any arbitrary axis, the probability of going one way or the other is the same. The ink will spread out according to a "Gaussian distribution" (assuming there is no barrier), and this is what we want to measure.

1.3   HOW TO MEASURE DIFFUSION?

1.3.1   We need gradient systems to measure the diffusion constant

In the previous section, we defined what diffusion is. We also learned that the only information we usually use for NMR/MRI is

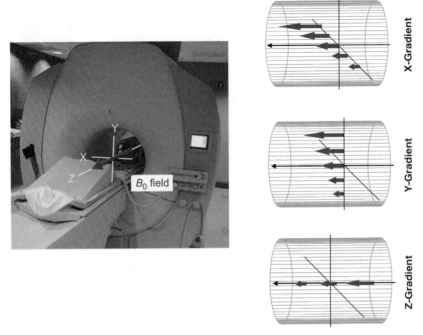

Fig. 1.4 The main magnetic field, B_0, and orientations of x, y, and z field gradient. The B_0 field, shown by the red arrows, lies along the magnet bore (z axis). The gradient units modulate the strength of the B_0 field linearly along one of the three axes.

signal intensity; peak height in NMR spectroscopy or pixel intensity in MRI. The intensity is dominated by water concentration (proton density). It is also influenced by signal relaxation time such as T_1 and T_2. Here, our task is to sensitize the signal intensity to the amount of water diffusion or diffusion constant. For this purpose, we need to use a special device called "pulsed magnetic field gradient" (referred to hereafter as simply "gradient").

The magnetic field gradient is a technology that introduces a linear magnetic field inhomogeneity (i.e., gradient) (Fig. 1.4).

In the magnet, the magnetic field, called the B_0 field, is pointing along the magnet bore, which is defined as the z axis. The x and y axes are the right–left and up–down orientations. When an x, y, or z gradient is applied, the B_0 strength is modulated linearly along each axis. Figure. 1.5 shows the relationship between the x, y, and z gradients and the anatomical axes. The field gradient along any arbitrary orientation can be created by combining the x, y, and z gradients.

As shown in Fig. 1.6, we can control strength and polarity (positive and negative) of the gradient. In addition, we can turn on and off gradients at any time point we want. Typically, we turn on gradients for only a short period of time (1–100 ms) and, thus, it is called "pulsed field gradient." In order to visualize schemes for gradient applications, we usually use diagrams such as that shown in Fig. 1.7.

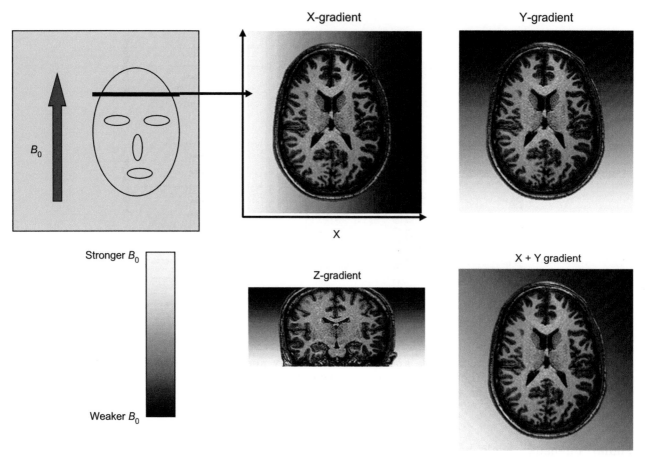

Fig. 1.5 Relationship between gradient orientations and brain anatomy. By combining the x and y gradients, an oblique angle can also be created, as shown in the diagram in the lower right corner.

1.3.2 Gradient pulses change signal frequency based on locations of water molecules

The frequency (ω) of the MR signal and B_0 field strength have a simple relationship:

$$\omega = \gamma B_0 \qquad (1.1)$$

where γ is the gyromagnetic ratio (γ of proton is 2.675×10^8 rad/s/T or 42.58 MHz/T). For example, if B_0 is 1.5 T, the signal frequency is 63.9 MHz. For both NMR spectroscopy and MRI, the magnetic field (B_0) is kept as homogeneous as possible, so that all water molecules give the same frequency (Fig. 1.8A). Once the field gradient is applied, water molecules at different locations start to resonate at different frequencies (Fig. 1.8B).[1]

Figure 1.9 shows a schematic to explain what happens to MR signals if a pair of positive and negative gradients is applied. After excitation RF pulse (time t_1), protons at different locations start to give MR signals at the same frequency. During the first gradient application (t_2), protons start to see different B_0 and resonate at different frequencies, depending on their locations. In this example, the red proton sees weaker B_0 and rotates slower, and *vice versa* for the blue proton. At the end of the gradient application and

[1] In reality, a sample tube (NMR) or each pixel (MRI) contains a large amount of water molecules, and only 0.003% (per Tesla) of them contribute to the signal. Therefore, the pink, green, and blue water in this example should be considered as ensembles of water molecules.

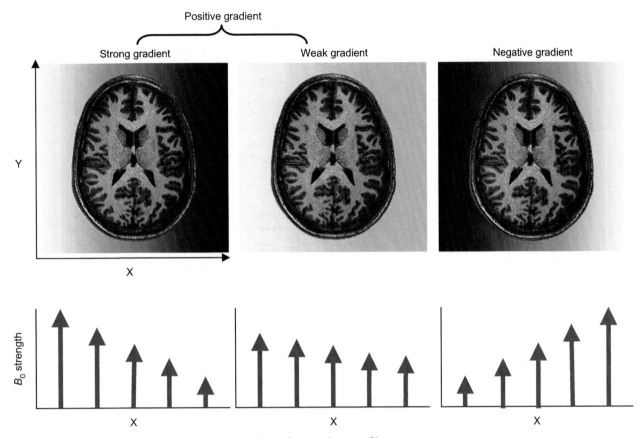

Fig. 1.6 Schematic diagram of strong, weak, and negative gradients.

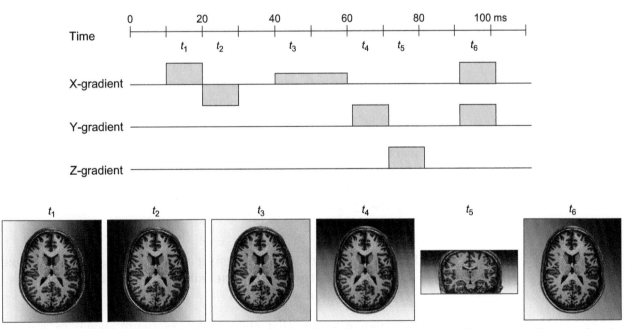

Fig. 1.7 An example of a gradient diagram. At time 10 ms, a strong positive x gradient pulse is applied for 10 ms, followed by a 10 ms pulse of negative strong x gradient. At time 40 ms, a weak long (20 ms) positive gradient pulse is applied. At time 60 and 70 ms, strong positive y and z gradients are applied for 10 ms, respectively. At time 90 ms, a gradient pulse is applied at 45° in the x–y plane for 10 ms.

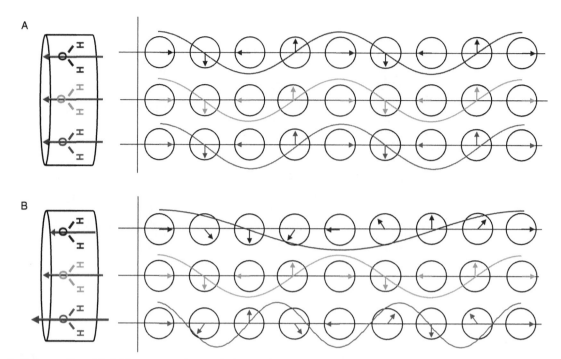

Fig. 1.8 Schematic of field homogeneity and signal frequency. (A) Usually, the magnetic field (red arrows) is kept as homogeneous as possible to ensure that water molecules at different locations see the same field strength and give the same signal frequency. (B) When a gradient is applied, each water molecule sees a different field strength, depending on its location. In this example, pink water sees a stronger field and resonates at a higher frequency, while blue water resonates at a lower frequency.

when the system regains the homogeneous B_0 (t_3), the phases (locations of the rotating arrows) of the signals are no longer identical among the protons. This leads to loss of overall signal (the bottom waveform), which is a sum of signals from all water molecules. Therefore, the first gradient is called the "dephasing" gradient. During t_3, all protons resonate at the same frequency, but their signal phases remain dephased. During the second gradient (t_4), because it has the opposite polarity, the red proton starts to rotate faster, catching up with other protons, and the blue one starts to rotate slower. If the strength and length of the second gradient are identical to those of the first one, the protons should regain the same phase at the end of the second gradient, as if nothing has happened in the past. Therefore, the second gradient is called "rephasing" gradient.

1.3.3 When a pair of dephasing and rephasing gradients are applied, the signal is sensitized to molecular motions (diffusion weighting)

Although it seems that the pair of gradients shown in Fig. 1.9 are not doing anything for the experiment, the resultant signal is now sensitized to diffusion and, thus, the signal is "diffusion-weighted," i.e., the gradient "diffusion-weights" the signal. This is because perfect refocusing happens only when water molecules

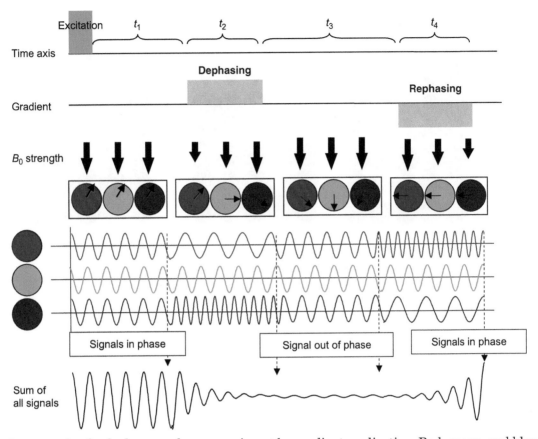

Fig. 1.9 An example of a dephase–rephase experiment by gradient application. Red, green, and blue circles indicate three water molecules located at different positions in a sample tube. Thick arrows indicate the strengths of magnetic field strength (B_0), and narrow arrows indicate phases of MR signals from each molecule.

do not change their locations in between the applications of the two dephase–rephase gradients. As shown in Fig. 1.10A, what the dephasing gradient does is to "tag" locations of water molecules using their signal phase. If water moves, it results in disruption of the phase gradient across the sample. After the rephrasing gradient, those molecules that have moved can, in principle, be detected because they have different phases from the other, stationary molecules. MR cannot measure the phase of individual water molecules, but it can detect the imperfect rephasing by the loss of signal intensity.

Here, several important points about diffusion measurement by MR should be made. First, it is noninvasive and does not require injection of any chemical tracers. Second, it measures the water motion along a predetermined axis. In this example, water moved along a horizontal axis (indicated by yellow boxes) can be detected because it perturbs the phase gradation. However, water motion along a vertical axis (indicated by a green box) does not have any effect and cannot be detected because no field gradient was applied in this direction. Third, diffusion and flow motions lead to different outcomes in this experiment. As shown in Fig. 1.10B, coherent motions such as flow or bulk motions of the subject result in perfect refocusing (thus, no signal loss) and shift of signal phase.

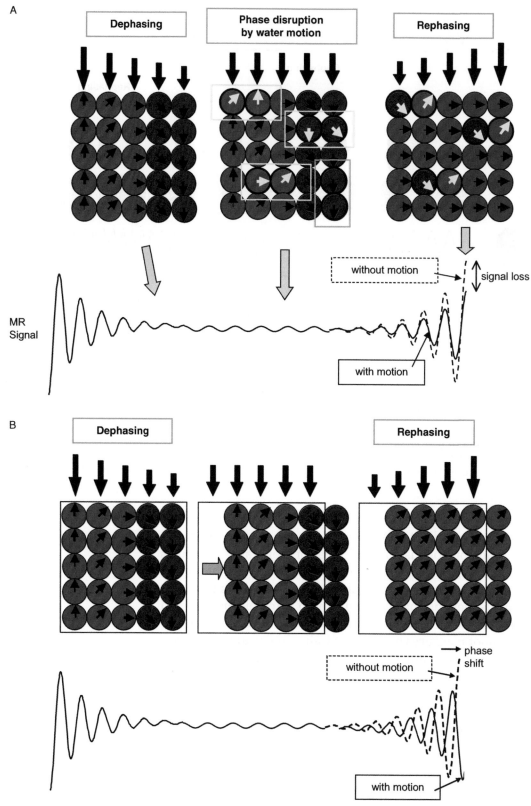

Fig. 1.10 Effects of molecular motions in an experiment with a pair of gradients. Large black arrows indicate the strength of magnetic field (B_0), and small arrows indicate phases of individual protons. The phase is also indicated by gradation of colors. (A) Diffusion process in which protons move randomly. Protons (circles) moved in between the two gradients are indicated by boxes. (B) Coherent motion of water such as flow or bulk motion.

Therefore, the incoherent motion (signal loss) and coherent motion (phase shift) can, in principle, be separated in the MR diffusion measurement. In practice, however, the flow can occur along multiple orientations within a pixel (such as small blood vessels), and bulk motion such as brain pulsation is not rigid and uniform as shown in Fig. 1.10B. In these cases, the coherent motion often leads to signal loss and interferes with the diffusion measurements. In a diffusion measurement, the size of the molecular displacement during the measurement is typically $1\text{--}20\,\mu\text{m}$, depending on sample, temperature, and pulse sequence, and the experiment is designed in such a way that the diffusion of this degree leads to a signal loss of typically 10–90% (see Chapter 2). In such experiments, water motions by blood flows are far larger, and the blood signal (about 5% of water population in the brain) should be completely dephased. The effects of bulk motions, such as brain pulsation and involuntary movement of the subject, are more troublesome. This issue will be discussed in more detail in Chapter 6.

Anatomy of diffusion measurement

2.1 A SET OF UNIPOLAR GRADIENTS AND SPIN-ECHO SEQUENCE IS MOST WIDELY USED FOR DIFFUSION WEIGHTING

In Fig. 1.9 of Chapter 1, it was shown that a pair of positive and negative gradients (called "bipolar" gradients) could be used to introduce diffusion weighting into the MR signal. One of the drawbacks of this approach is that we lose a fair amount of signal due to T_2^* decay between the two gradient pulses (typically 20–40 ms in clinical scanners). To refocus the T_2^* decay, it is more common to use a spin-echo sequence, in which the signal decays by T_2 and we lose less signal during diffusion weighting. The spin-echo sequence employs a 180° refocusing radio frequency (RF) pulse that reverses the signal phase. In this case, the rephasing gradient has to have the same sign as the dephasing gradient (Fig. 2.1B). Because both gradients have the same sign, they are called "unipolar" gradients. For both bipolar and unipolar approaches, the experiment is sensitive to molecular diffusion when the signal phase is being dephased (time period indicated by red lines).

In Fig. 2.2, some examples of other types of diffusion-weighting sequences are shown. In Fig. 2.2A, the polarity of the bipolar pair is inverted. This simply inverts the sign of dephasing, which leads to the same results as Fig. 2.1A. The sequence shown in Fig. 2.2B is sometimes used for special types of diffusion-weighting experiments. Figure 2.2C shows a widely used sequence because it compensates for the imperfection of gradient instruments. The amount of diffusion weighting is related to the area under the red lines; the larger the area, the more the diffusion weighting. The sequences shown in Fig. 2.2A and Fig. 2.2C thus have more weighting than those in Fig. 2.2B.

2.2 THERE ARE FOUR PARAMETERS THAT AFFECT THE AMOUNT OF SIGNAL LOSS

We have learned that MR signal intensity becomes sensitized to molecular diffusion by applying a pair of dephasing and rephasing gradients because movement of molecules between the two gradient pulses leads to imperfect rephasing and signal loss. Let us now define S_0 as the signal intensity without gradients, and S as the signal with a pair of gradients. What are the parameters that

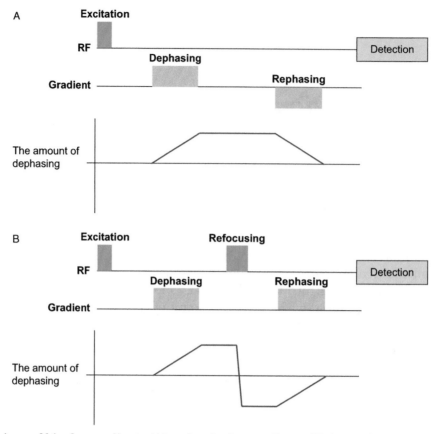

Fig. 2.1 Comparison of bipolar gradients (A) and unipolar gradients (B) in a spin-echo sequence. Red lines show the amount of signal dephasing introduced by the first gradient.

affect the signal attenuation (S/S_0)? Apparently, the longer the interval between the two pulses ($\varDelta$ in Fig. 2.3), there is more chance for the water molecules to move around, which leads to more signal loss. The higher the diffusion constant (D), the greater the chance for waters to change their position within a fixed amount of time, $\varDelta$. Another parameter is the amount of initial dephasing, which is decided by the area (strength (G) × length (δ)) of gradients. These are the four parameters that dictate the amount of signal loss ($S/S_0 = f(\varDelta, D, G, \delta)$).

The reason why $G \times \delta$ (or the amount of initial dephasing) influences the signal attenuation may not be intuitive at first glance. This is explained in Fig. 2.4. Among these four parameters, G, $\varDelta$, and δ are experimental parameters that we can control. By changing them, we can control the amount of signal loss ("diffusion weighting"), and the pair of gradients is called "diffusion-weighting gradients." The signal intensities with and without diffusion-weighting gradient pulses (S and S_0) are the experimental results that we measure, and since G, δ, and $\varDelta$ are all known, it is possible to calculate D, the diffusion constant that we are interested in.

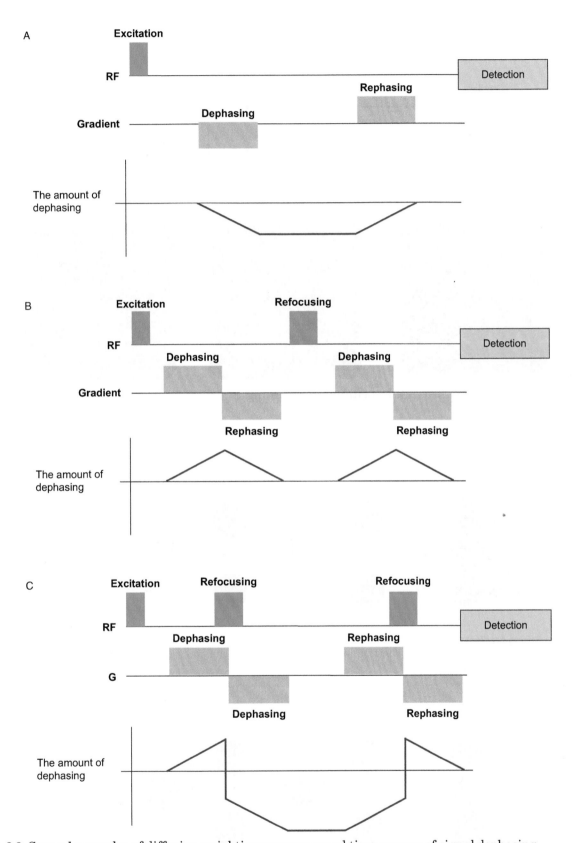

Fig. 2.2 Several examples of diffusion-weighting sequences and time courses of signal dephasing.

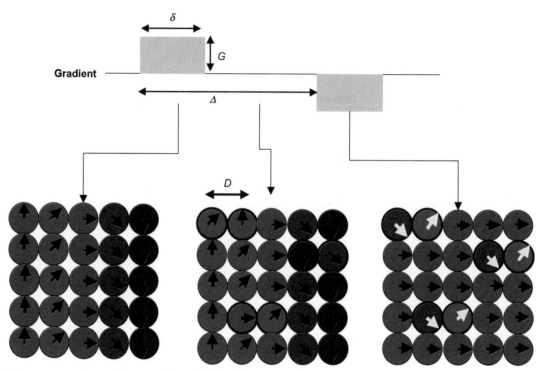

Fig. 2.3 Parameters that affect results of diffusion weighting.

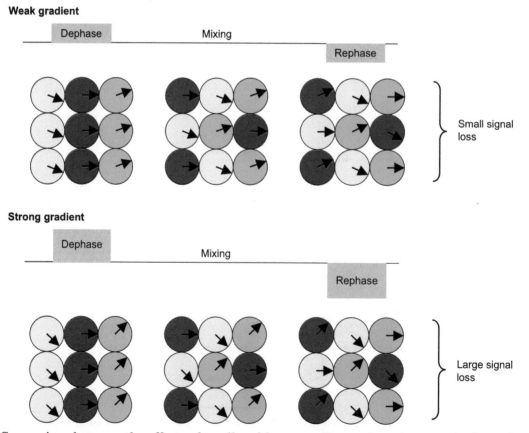

Fig. 2.4 Comparison between the effects of small and large gradient pulses. A strong (or longer) gradient leads to more initial dephasing along the gradient axis. The same amount of molecular motion leads to more signal loss and, thus, more diffusion weighting. In this figure, water molecules are indicated by yellow, pink, and cyan to visualize changes in their locations.

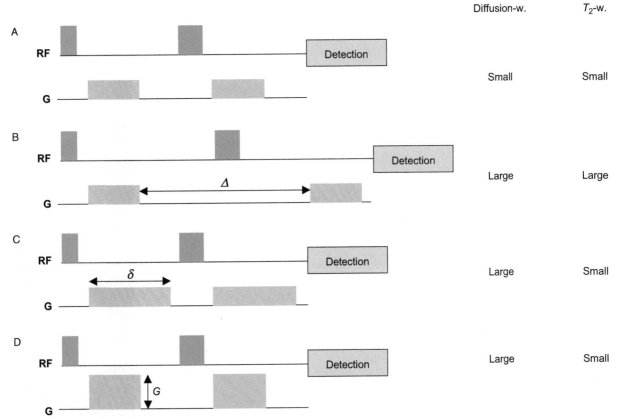

Fig. 2.5 Several ways of achieving different diffusion weighting. Diagram (A) shows a spin-echo sequence with weak diffusion weighting. Diffusion weighting can be increased by lengthening Δ (B) or δ (C) or by applying a stronger gradient G (D). Each experiment leads to different diffusion-weighting and T_2 weighting.

2.3   THERE ARE SEVERAL WAYS OF ACHIEVING A DIFFERENT DEGREE OF DIFFUSION WEIGHTING

As described in the previous section, we are interested in signal loss due to the application of diffusion-weighting gradients. To observe the signal loss, we have to compare results from at least two experiments: one without (or with weak) diffusion-weighting gradients and the other with stronger weighting. To change the amount of diffusion weighting, we have three parameters that we can control: G, δ, and Δ, as shown in Fig. 2.5. Among these approaches, changing the gradient separation (Δ) is not ideal because it changes echo time and, thus, the resultant signal loss would be influenced by both diffusion and T_2 weighting (Fig. 2.5B). Lengthening δ (Fig. 2.5C) works but we usually do not have much room between the excitation and refocusing RF pulses to accommodate longer δ. The most common way to increase diffusion weighting is to increase gradient strength, G, as shown in Fig. 2.5D. In this way, we can keep the effects of proton density and relaxation times (T_1 and T_2) unchanged, and the resultant signal loss would be solely due to the diffusion process.

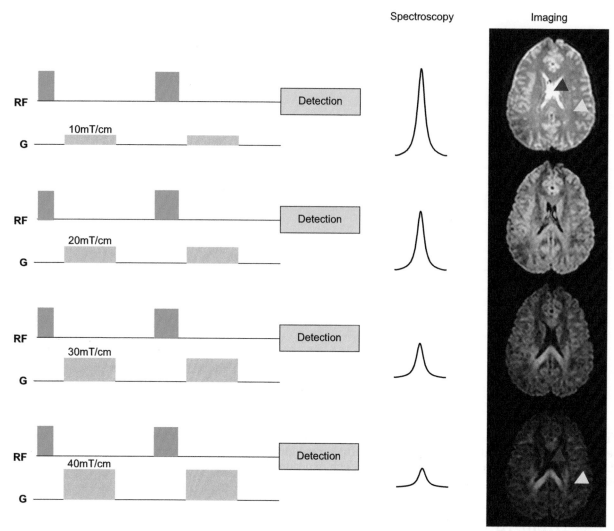

Fig. 2.6 A typical spin-echo-based diffusion measurement, in which a series of measurements with different gradient strengths are made. Depending on the detection scheme, we can carry out either spectroscopy or imaging. From the images, we can clearly see that the intensity of the ventricle (indicated by red arrowheads) lose more intensity than the parenchyma (yellow arrowheads), indicating faster diffusion constants.

An example of typical diffusion constant measurement is shown in Fig. 2.6. In this example, four measurements are performed with gradient strengths of 10, 20, 30, and 40 mT/cm. We can refer to all the images shown in Fig. 2.6 as 'diffusion-weighted images' because gradients are applied in all cases. However, the one with 40 mT/cm shows the typical contrast of a heavily diffusion-weighted image, in which regions with a high diffusion constant (e.g., ventricle indicated by a red arrow) have low signal intensity and those with a low diffusion constant (indicated by a yellow arrow) have high signal intensity.

Chapter 3

Mathematics of diffusion measurement

3.1 WE NEED TO CALCULATE DISTRIBUTION OF SIGNAL PHASES BY MOLECULAR MOTION

(Those who are not interested in the mathematical derivation may proceed to Section 3.2.)

First, we need to determine how a phase distribution in space is introduced by a gradient pulse with a strength of G and a duration of δ (Fig. 3.1). The amount of phase difference with respect to an arbitrarily defined reference point ($x = 0$) is:

$$\phi(x) = e^{i\gamma G\delta x} \tag{3.1}$$

where γ is the gyromagnetic ratio (2.765×10^8/s, T) and x is the distance from the reference point. If we apply 40 mT/m gradient for 20 ms, the $\gamma G\delta$ term becomes 2.212×10^2 s/mm. The term x is the location along the x-axis. The diffusion constant (D) of water inside the brain is about 1.0×10^{-3} mm^2/s and the period between the two gradient pulses (Δ) is, say, 30 ms (this is called diffusion time). Using Einstein's equation $\sigma = \sqrt{2Dt}$, where σ is the mean (average) diffusion distance, D the diffusion constant, and t the diffusion time, we can estimate that water moves approximately 8 μm on average. If we substitute $x = 8$ μm, the $\gamma G\delta x$ term becomes 1.7. This is about 1/4 of 2π. This means that the phase of water signal is about 90° off between two places separated by 8 μm. This is the condition after the first gradient pulse. Note that at this point water molecules have not moved yet.

After the phase gradient is introduced across the sample, water molecules start to move during the time period, Δ (Fig. 3.2). In this figure, we pay attention to four water molecules (painted red) at the same initial position. Although the water molecules move in 3D space, we are interested only in the movement along the gradient axis, which is, in this case, horizontal orientation (x-axis). After the time period Δ and a rephasing pulse, each water molecule gets a phase shift that is proportional to the amount of movement. The further the water moves, the more phase shift it gets (Eq. 3.1). Please note that diffusion is a random process, therefore, the probability of the water molecules moving right or left is the same.

Figure 3.2 shows several important facts. First, we can control the gradient scheme (G, δ) to introduce any desired amount of phase gradient at the time point, t_2. Second, the amount of phase shift after the second pulse (time point t_4) can be estimated from

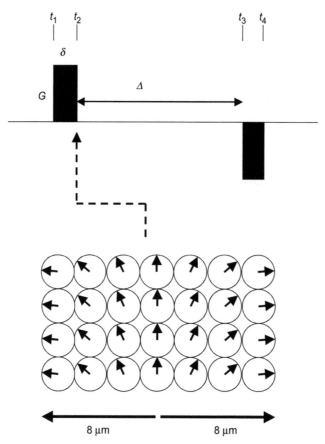

Fig. 3.1 Gradation of signal phase introduced by the first gradient pulse (time point t_2). With $G = 40\,\mathrm{mT/m}$ and $\delta = 20\,\mathrm{ms}$, there is about $90°$ phase difference between locations $8\,\mu\mathrm{m}$ apart.

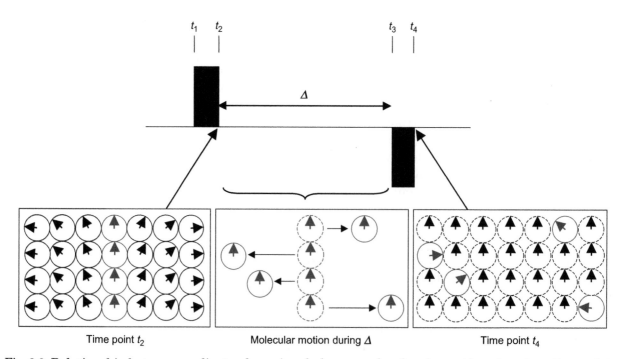

Fig. 3.2 Relationship between gradient pulses, signal phases, and molecular motion at various time points. After the time point t_2, only the four molecules painted red are followed. If all the other molecules do not move, their signal phases refocus at time point t_4.

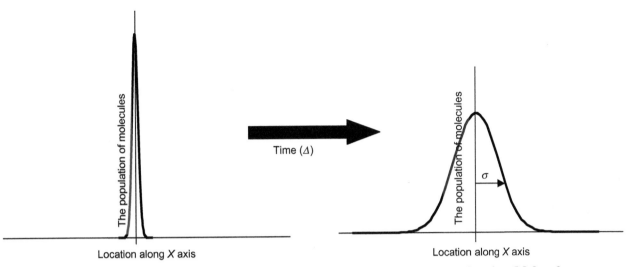

Fig. 3.3 Molecular motion obeys Gaussian distribution for freely diffusing molecules. Molecules concentrated in one place (e.g., red-colored water molecules in Fig. 3.2) start to spread out as time elapses. The distribution (Gaussian function) gives the population of molecules after time t ($= \varDelta$ for the gradient scheme used in Fig. 3.1) at each location along the x-axis. The width (σ) is a function of diffusion constant (D) and time ($\varDelta$) ($\sigma = \sqrt{2D\varDelta}$).

Eq. 3.1 by plugging numbers for G, δ, and x. Third, the only missing piece that we have not considered is how water molecules move during $\varDelta$. Previously, we have seen that if $\varDelta$ is 30 ms, the water molecules move by 8 μm. However, this is an average value, and it is not that all water molecules move 8 μm. Therefore, the next important task is to characterize the amount of movement. In Fig. 3.2, we use only four water molecules as an example, but in a real situation, there are many more molecules within a pixel. If we can assume free diffusion, distribution of water molecules can be described by Gaussian distribution (Fig. 3.3). In general, the Gaussian function is described as:

$$\frac{1}{\sigma\sqrt{2\pi}}\mathrm{e}^{-x^2/2\sigma^2} \qquad (3.2)$$

where $1/\sigma\sqrt{2\pi}$ is the scaling factor to normalize the area under the curve to 1. This gives us the population of water molecules at location x. The parameter σ controls the width of the curve. In our case, this width tells us how far water molecules travel on average. As mentioned earlier, we can use Einstein's equation, $\sqrt{2Dt}$, to estimate the average distance water travels. So, by substituting σ with $\sqrt{2Dt}$, we can obtain:

$$P(x, t) = \frac{1}{\sqrt{4\pi\ Dt}}\mathrm{e}^{-x^2/4Dt} \qquad (3.3)$$

where $P(x,t)$ tells population of water at location x at time point t. This is a function of t (or $\varDelta$, in our case), x (when we use the x-gradient), and D. The longer the t ($= \varDelta$), the wider the distribution becomes. For the fixed length of $\varDelta$, higher diffusion constants (D) lead to wider distribution (Fig. 3.4).

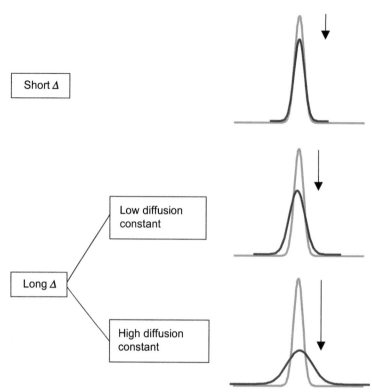

Fig. 3.4 Distribution of water molecules (width of Gaussian) depends on time and diffusion constant. The longer we wait (Δ) or the higher the diffusion constant, the more spreading occurs. The blue line represents water molecules in a specific location ($x = 0$) at time t_1, and red represents distribution at time t_4 in Fig. 3.2. The heights of the peaks indicate the number of water molecules (population) and the horizontal axes are distance from $x = 0$.

We have now mathematically characterized three important processes to calculate the amount of signal attenuation (S/S_0): (1) the average distance traveled by water molecule (σ) with diffusion constant (D) and time Δ (Einstein's equation, $\sigma = \sqrt{2Dt}$, where $t = \Delta$); (2) the amount of phase shift that is decided by applied gradient ($G\delta$) and traveled distance (x) of water (Eq. 3.1); and (3) the population of water molecules at location x after time t and with diffusion constant D (Eq. 3.3). Our final task is to combine all these equations to calculate the amount of signal attenuation, which should be a function of t, D, and applied gradient ($G\delta$) (Fig. 3.5). As shown in Fig. 3.6, the total signal can be calculated by adding all the signal phases that are weighted by population at each location. When no gradient is applied, there is no dispersion in the signal phase and all signals add constructively (zero $G\delta$ in Fig. 3.5). When there is more phase dispersion (larger $G\delta$) or more population dispersion (wider width (σ) of Gaussian distribution), there would be more signal loss.

It is intuitive that the total signal can be calculated by summing up the product of population and signal phase along location x; signal $= \sum_x P(x, \Delta) \ \phi(x)$. This concept is explained in Fig. 3.6, in which populations of molecules with specific phase positions are classified discretely. In a real situation, however, both phase and population dispersions are continuous. So, the equation becomes

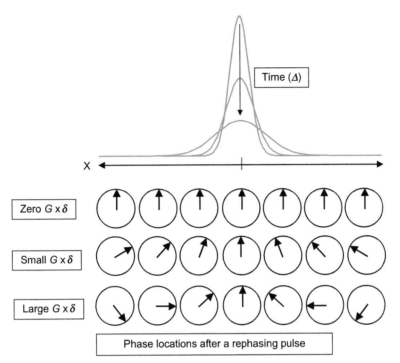

Fig. 3.5 Relationship between the population of moved molecules and their phases at time t_4 in Fig. 3.2. The amount of phase shift after a refocusing pulse is determined by $G\delta$ and distance traveled by molecules, x (compare the phase positions of moved molecules with the diagram in Fig. 3.2). The population of water molecules at each location obeys Gaussian distribution, which is determined by time (Δ) and diffusion constant (D).

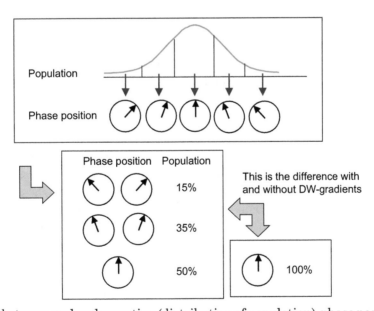

Fig. 3.6 Relationship between molecular motion (distribution of population), phase position, and signal loss.

(for readers who are not interested in exact derivation, jump to Eq. 3.18):

$$\text{Signal} = \int_x P(x, t)\phi(x)\mathrm{d}x = \frac{1}{\sqrt{4\pi \, D \, \Delta}} \int_x e^{-x^2/4D\Delta} e^{i\gamma G\delta x}\mathrm{d}x, \quad (3.4)$$

for molecules with diffusion constant D and an experiment with diffusion time (t) of Δ and applied gradients of $G\delta$. The gradient is assumed to be applied along the x-axis. Note that if we do not apply gradient $G = 0$, signal phase ($\phi(x) = e^{i\gamma G\delta x}$) becomes 1. Therefore, this term drops out and signal becomes:

$$\text{Signal} = \int_x P(x, t)\mathrm{d}x = \frac{1}{\sqrt{4\pi\ D\Delta}}\int_x e^{-x^2/4D\Delta}\mathrm{d}x = 1 \qquad (3.5)$$

This is simply an integration of the Gaussian curve and, as explained in Eq. 3.2, the solution (signal intensity) is 1. Because the phase term (ϕ) is always less than 1, application of gradients ($G \neq 0$) leads to a signal less than 1. Eq. 3.4 can be calculated as follows:

$$\frac{1}{\sqrt{4\pi D\Delta}}\int_x e^{-x^2/4D\Delta}e^{i\gamma G\delta x}\mathrm{d}x$$

$$= \frac{1}{\sqrt{4\pi D\Delta}}\left[\int_x e^{-x^2/4D\Delta}\cos(\gamma G\delta x)\mathrm{d}x - i\int_x e^{-x^2/4D\Delta}\sin(\gamma G\delta x)\mathrm{d}x\right]$$

$$(3.6)$$

The second imaginary term is 0 because it is an asymmetric function and integration over $-\infty$ to ∞ becomes 0. The integration of the real terms is:

$$\text{Signal} = e^{-\gamma^2 G^2\delta^2 D\Delta} \qquad (3.7)$$

This signal intensity is normalized, so that its maximum value is 1 when $G = 0$ (no diffusion weighting). In a practical situation, we have a certain number for signal intensity read from the MRI scanner. If we assign S_0 and S for the signal intensity without and with diffusion weighting, Eq. 3.7 becomes:

$$S = S_0 e^{-\gamma^2 G^2\delta^2 D\Delta} \qquad (3.8)$$

When deriving Eq. 3.8, we operate under the assumption that the gradient pulses are applied instantaneously so that diffusion of water starts from time point t_2 and finishes at t_3 in Fig. 3.2. This is only true when we can assume a so-called short gradient limit, in which we can neglect the diffusion process between time t_1 and t_2 and between t_3 and t_4, as shown Fig. 3.7A ($\delta \ll \Delta$). In a practical situation, the gradient length (δ) is usually long (5–30 ms), and we cannot neglect molecular motion during the gradient pulse. So, where did we cheat during the derivation? It was in Eq. 3.4, in which we substituted variable t with a constant Δ (for time between t_2 and t_3). We also treated ϕ as a function of only x. For more practical experiments, such as those shown in Fig. 3.7B and Fig. 3.7C, applied gradient areas ($G\delta$) and gradient strength (G) are time variant and, therefore, the introduced phase gradation (Fig. 3.1) is now a function of both location (x) and time (t):

$$\phi(x, t) = e^{-i\gamma G(t)tx} \qquad (3.9)$$

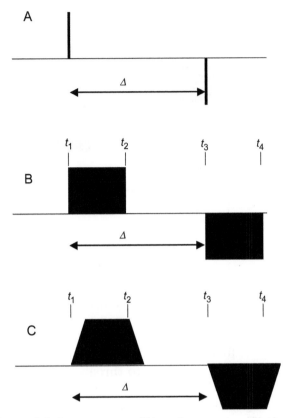

Fig. 3.7 A simplified diffusion-weighting sequence (A) and a more realistic sequence (B) and (C). In (A), dephasing and rephasing gradients are applied instantaneously, and diffusion during the gradient pulses is negligible (narrow gradient limit).

where $G(t)$ is the shape of the gradient pulse (a step function in Fig. 3.7B and a trapezoidal in Fig. 3.7C). The Gaussian distribution term $P(x, t)$ remains Eq. 3.3, in which t is not substituted with Δ. Eq. 3.4 then becomes:

$$\int_x P(x, t)\ \phi(x, t)\mathrm{d}x = \frac{1}{\sqrt{4\pi\ Dt}} \int_x e^{-x^2/4Dt} e^{i\gamma G(t)tx} \mathrm{d}x \qquad (3.10)$$

To calculate signal intensity, we need to integrate not only location (x) but also time (t). After integration of x, the equation becomes:

$$S = S_0 e^{-D\gamma^2 \int_{t_1}^{t_4} (\int_0^t G(t')\mathrm{d}t')^2 \mathrm{d}t} \quad \text{or} \quad \ln\left(\frac{S}{S_0}\right) = -D\gamma^2 \int_{t_1}^{t_4} \left(\int_0^t G(t')\mathrm{d}t'\right)^2 \mathrm{d}t$$

$$(3.11)$$

Note the double integral for the time variables. The first integral integrates over t' which starts from the application of the first gradient and ends at the time variable t. The second integral then integrates time from t_1 to t_4. Let us follow the equation using the gradient scheme in Fig. 3.7B. The time course of the function $f(t) = \int_0^t G(t')\mathrm{d}t'$ is shown in Fig. 3.8.

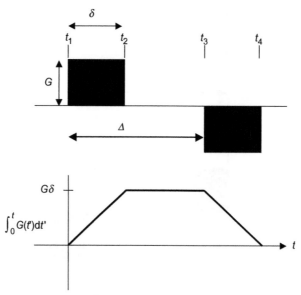

Fig. 3.8 The time course of the function $\int_0^t G(t')dt'$, which describes the area of applied gradient and is related to the amount of signal phase gradation (ϕ).

To integrate this function over time period t_1–t_4, we need to divide the time into three segments, t_1:t_2, t_2:t_3, and t_3:t_4, and integrate them separately. For time period t_1:t_2, $\int_0^t G(t')dt'$ is Gt; for t_2:t_3, it is a constant $G\delta$, and for t_3:t_4, it is $G\delta -G(t-t_3)$. Then Eq. 3.11 becomes:

$$\ln\left(\frac{S}{S_0}\right) = -D\gamma^2\left(\int_{t_1}^{t_2}G^2t^2dt + \int_{t_2}^{t_3}G^2\delta^2dt + \int_{t_3}^{t_4}\left(G\delta - G(t-t_3)\right)^2dt\right)$$

(3.12)

We can simplify this equation using some substitution and time definition. First, let us define time t_1 as our origin of time axis; $t_1 = 0$. Then time $t_2 = \delta$. Second, let us define the time interval t_1:$t_3 = \Delta$ as shown in Fig. 3.7 and Fig. 3.8. Then time $t_4 = \Delta + \delta$.

The first term $\int_{t_1}^{t_2} G^2t^2dt$ then becomes:

$$\int_0^\delta G^2t^2dt = \frac{1}{3}G^2\delta^3$$

(3.13)

The second term $\int_{t_1}^{t_2} G^2\delta^2dt$ becomes:

$$\int_\delta^\Delta G^2\delta^2dt = G^2\delta^2(\Delta - \delta) = G^2\delta^2\Delta - G^2\delta^3$$

(3.14)

The third term $\int_{t_3}^{t_4}(G\delta - G(t-t_3))^2dt$ becomes:

$$\int_\Delta^{\Delta+\delta} (G\delta - G(t-\Delta))^2dt = \frac{1}{3}G^2\delta^3$$

(3.15)

By adding all three terms:

$$\ln\left(\frac{S}{S_0}\right) = -\gamma^2G^2\delta^2\left(\Delta - \frac{\delta}{3}\right)D$$

(3.16)

Interestingly, the exact solution for the trapezoidal scheme (Fig. 3.7C) is identical to Eq. 3.16 as long as the time period $\varDelta$ is defined as shown in Fig. 3.7C. The reader is challenged to derive it following Eq. 3.11 to 3.16. [Hint: In this case, however, the time period needs to be divided into more sections.]

3.2 SIMPLE EXPONENTIAL DECAY DESCRIBES SIGNAL LOSS BY DIFFUSION WEIGHTING

Equation 3.16 can be written in many different forms:

$$\frac{S}{S_0} = e^{-\gamma^2 G^2 \delta^2 (\varDelta - \delta/3)D},$$

$$S = S_0 e^{-\gamma^2 G^2 \delta^2 (\varDelta - \delta/3)D},$$

$$\ln\left(\frac{S}{S_0}\right) = -\gamma^2 G^2 \delta^2 (\varDelta - \delta/3)D,$$

$$\ln(S) = \ln(S_0) - \gamma^2 G^2 \delta^2 (\varDelta - \delta/3)D$$

(3.17)

These are different expressions of the same equation. As described earlier, S and S_0 are signal intensities with and without diffusion weighting, respectively. The parameter γ is a gyromagnetic ratio (2.765×10^8 rad/s, T). The parameters G, δ, and $\varDelta$ can be controlled by us. These parameters are often abbreviated to one parameter, b ($= \gamma^2 G^2 \delta^2 (\varDelta - \delta/3)$). So, the preceding equations can be simplified to:

$$\ln(S) = \ln(S_0) - bD$$

$$S = S_0 e^{-bD}$$

(3.18)

This equation is similar to a simple linear equation, $Y = \text{constant} - a\,X$, where X is an independent variable corresponding to our "b" value, constant is $\ln(S_0)$, and Y is a dependent variable corresponding to our measurement results, $\ln(S)$. If we plot X versus Y (meaning b versus S), we should observe linear signal decay, as shown in Fig. 3.9. In this equation, there are two unknowns, S_0 and D. Therefore, we need at least two measurement results to solve it.

3.3 DIFFUSION CONSTANT CAN BE OBTAINED FROM THE AMOUNT OF SIGNAL LOSS BUT NOT FROM THE SIGNAL INTENSITY

In Fig. 3.10, the results of diffusion imaging shown in Fig. 2.6 (Chapter 2) are copied. To investigate water diffusion at each pixel, we have to study the intensity of each pixel individually. It must be stressed that the diffusion constant cannot be obtained from the signal intensity of one image. Let us consider pink and blue pixels in an image with $b = 100\,\text{s/mm}^2$. In this example, we can find that pixel intensity of the pink pixel is

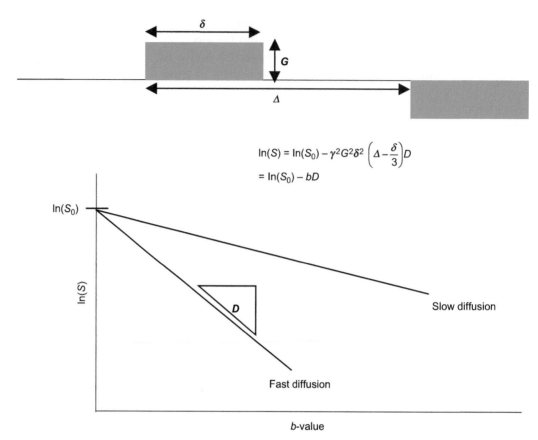

$$\ln(S) = \ln(S_0) - \gamma^2 G^2 \delta^2 \left(\Delta - \frac{\delta}{3} \right) D$$
$$= \ln(S_0) - bD$$

Fig. 3.9 Relationship between gradient parameters G, δ, Δ, and signal intensity S.

lower than that of the blue pixel. However, we cannot conclude that the diffusion constant is higher in the pink pixel. This is because the intensity of each pixel is weighted by proton density, T_1, T_2, and diffusion properties of local water molecules. Therefore, the intensity difference between pink and blue pixels (indicated by an asterisk) cannot indicate if there is any difference in diffusion constants in these regions. Suppose we use a very long repetition time, and T_1 weighting is negligible; we can write an equation of signal intensity of the pink and blue pixels as follows:

$$S_{b=100,\text{pink}} = PD_{\text{pink}} e^{-1/T_{2,\text{pink}} TE} e^{-b_{100} D_{\text{pink}}} = S_{0,\text{pink}} e^{-b_{100} D_{\text{pink}}}$$
$$S_{b=100,\text{blue}} = PD_{\text{blue}} e^{-1/T_{2,\text{blue}} TE} e^{-b_{100} D_{\text{blue}}} = S_{0,\text{blue}} e^{-b_{100} D_{\text{blue}}}$$
$$(3.19)$$

where PD stands for proton density and TE for echo time. The term $PD e^{-1/T_2 TE}$ is simplified to S_0. If we find $S_{\text{pink}} < S_{\text{blue}}$, we cannot decide whether this is due to $S_{0,\text{pink}} < S_{0,\text{blue}}$ (difference in proton density and/or T_2) or due to $e^{-bD\text{pink}} < e^{-bD\text{blue}}$ (difference in diffusion constant).

When we apply heavy diffusion weighting (e.g., the image with $b = 1394$ s/mm^2), we get another set of signal intensities:

$$S_{b=1394,\text{pink}} = S_{0,\text{pink}} e^{-b_{1394} D_{\text{pink}}}$$
$$S_{b=1394,\text{blue}} = S_{0,\text{blue}} e^{-b_{1394} D_{\text{blue}}}$$
$$(3.20)$$

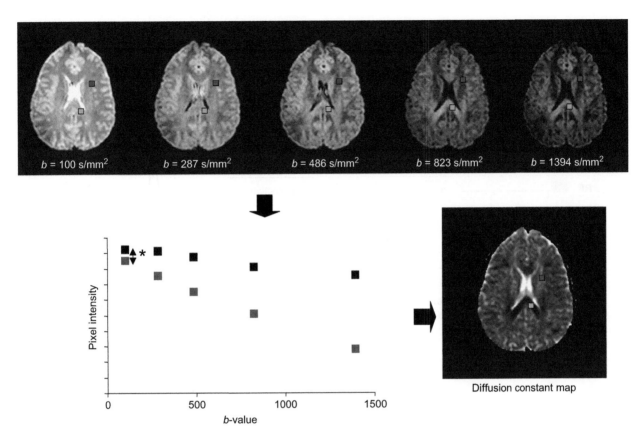

Fig. 3.10 Relationship between gradient strength and signal intensity. As the gradient strength increases, so does diffusion weighting. This leads to signal loss, and the slope, but not the absolute signal intensity, carries information about the diffusion constant. Once the diffusion constant is determined from the slope at each pixel, a so-called diffusion constant map can be obtained, in which brightness is proportional to diffusion constants; the brighter it is, the higher the diffusion constant. In this example, the pink pixel has a large amount of signal loss, suggesting fast diffusion. In the diffusion constant map, this region becomes bright. The blue pixel, on the other hand, has much smaller signal loss and a shallow slope. The diffusion constant in this region is slow and becomes darker in the diffusion constant map. Data courtesy of Jonathan Farrell and Peter van Zijl.

With this heavy diffusion weighting, it is likely that signal intensity is dominated by diffusion weighting. Therefore, we may say that the dark regions (pink pixel) in this image have a higher diffusion constant. However, a more accurate way to determine the diffusion constant is based on the amount of signal decay;

$$S_{b=1394,\text{pink}}/S_{b=100,\text{pink}} = e^{-(1394-100)D_{\text{pink}}}$$
$$S_{b=1394,\text{blue}}/S_{b=100,\text{blue}} = e^{-(1394-100)D_{\text{blue}}}$$

$$(3.21)$$

By taking the ratio of the two experiment results, terms for PD and T_2 drop out and the equation becomes independent of them. S$_0$, as long as we keep the echo time (TE) constant and observe the amount of signal loss, we can obtain diffusion constant (D) independent of PD and T_2.

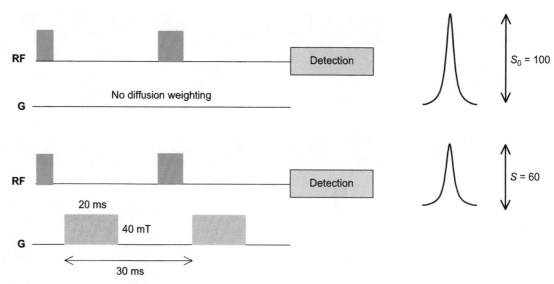

Fig. 3.11 An example of a diffusion constant measurement. The first experiment is without diffusion weighting, which provides $S_0 = 100$. The second experiment is diffusion weighted by $b = 1141\,s/mm^2$ and signal $S = 60$.

3.4   FROM TWO MEASUREMENTS, WE CAN OBTAIN A DIFFUSION CONSTANT

A hypothetical experiment is shown in Fig. 3.11; with no diffusion weighting, signal intensity $S_0 = 100$. In the second experiment, diffusion weighting is applied by using a pair of gradients with $G = 40\,mT/m$, $\delta = 20\,ms$, and $\Delta = 30\,ms$. This amounts to $b = 1141\,s/mm^2$ and the resultant signal intensity $S = 60$. By plugging these numbers into Eq. 3.18, we can obtain a diffusion constant D:

$$\ln\left(\frac{60}{100}\right) = -1141D$$

$$D = 0.45 \times 10^{-3}\,mm^2/s$$

Sometimes, it may be difficult to obtain S_0. In fact, it is usually impossible not to use gradient pulse in imaging. This means that we always have a finite amount of diffusion weighting in all imaging sequences. When the b-value of required gradient pulses is negligibly small, we can use such data as S_0. Fortunately, S_0 (= non–diffusion weighting) data is often not necessary in determining D. We are interested in quantifying the slope of signal decay; hence, as long as we have two measurement points at different b values, we can calculate it. For example, suppose we do two experiments with 10 and 40 mT/m gradients, as shown in Fig. 3.12, at the respective b-values of 70 and 1141 s/mm². After the experiments, we get two diffusion-weighted signals, S_1 and S_2:

$$\ln(S_1) = \ln(S_0) - b_1 D \ \text{ or } \ S_1 = S_0 e^{-b_1 D}$$
$$\ln(S_2) = \ln(S_0) - b_2 D \ \text{ or } \ S_2 = S_0 e^{-b_2 D}$$

(3.22)

where S_1, S_2, b_1, and b_2 are 80, 60, 70 and 1141 s/mm², respectively.

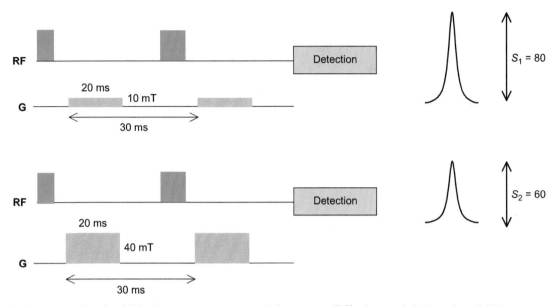

Fig. 3.12 An example of a diffusion measurement without non-diffusion-weighting signal (S_0).

TABLE 3.1

An example of the diffusion constant measurement of water

G (mT/m)	δ (ms)	Δ (ms)	B (s/mm²)	S (no unit)	ln (S)	D (mm²/s)*
30	5	10	14.3	11011	9.31	—
50	5	10	39.8	10621	9.27	1.41
80	5	10	101.9	9401	9.14	1.80
110	5	10	192.7	7703	8.95	2.00
140	5	10	312.2	6213	8.73	1.92
170	5	10	458.5	4420	8.39	2.05
200	5	10	634.6	3201	8.07	1.99

*Diffusion constant was measured from two points, using the results with 30 mT/m as the first point.

If we take a ratio of two experiments:

$$\frac{S_2}{S_1} = e^{-(b_2 - b_1)D} \quad \text{or} \quad \ln(S_2) = \ln(S_1) - (b_2 - b_1)D \qquad (3.23)$$

Solving this equation, we can obtain $D = 0.27 \times 10^{-3}$ mm²/s.

3.5 IF THERE ARE MORE THAN TWO MEASUREMENT POINTS, LINEAR LEAST-SQUARE FITTING IS USED

It is now clear that we can determine the diffusion constant if we have two measurement points. However, it is common to perform more than two measurements to improve the signal-to-noise ratio, as shown in Fig. 3.10. For example, Table 3.1 shows the result of the diffusion constant measurement of water.

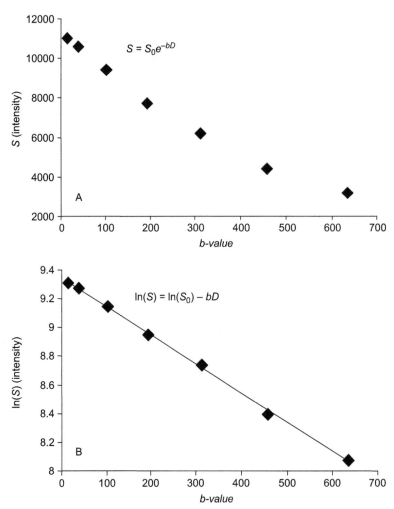

Fig. 3.13 Plotting of the results in Table 3.1. The results are plotted in the natural scale (A) and logarithmic (B) for the signal intensity (S). For (B), a linear-fitting result is shown, which was $Y = 9.35 - 0.00202X$.

In this experiment, measurements are made with seven different b values. The results are plotted in Fig. 3.13A. As shown in Table 3.1, we can pick up two measurement points and calculate a diffusion constant using Eq. 3.23. To use all seven measurement results to obtain a more accurate result, we need to use a fitting routine. We can use nonlinear fitting to fit a curve in Fig. 3.13A, but we can simplify the process by taking a log scale of signal intensity (Eq. 3.18). As shown in Fig. 3.13B, we can use a simple linear fitting such as least-square fitting. The dataset in Table 3.1 yields $Y = 9.35 - 0.00202X$ as a fitting result, where 9.35 corresponds to S_0 and 2.02×10^{-3} to the diffusion constant (mm^2/s).

Chapter 4

Principle of diffusion tensor imaging

4.1 NMR/MRI CAN MEASURE DIFFUSION CONSTANTS ALONG AN ARBITRARY AXIS

One of the most important features of the diffusion measurement by NMR/MRI is that it always measures diffusion along one predetermined axis. This can be understood from Fig. 4.1. When a gradient is applied along the horizontal axis, the signal becomes sensitive only to horizontal motion. Similarly, vertical motion can be detected if a vertical gradient is applied. As shown in Chapter 1, Fig. 1.5, we can apply a diffusion-weighting gradient along any arbitrary angle.

This unique feature of NMR/MRI may not be important if we are measuring free water diffusion inside a sample tube, because we expect the same diffusion constant regardless of measurement orientation. However, when we measure water diffusion inside a living system, we often find that the diffusion process has directionality.

4.2 DIFFUSION SOMETIMES HAS DIRECTIONALITY

Diffusion sometimes has directionality. This comment may sound as if it contradicts what was illustrated in Fig. 1.3, namely, we are not interested in flow that has directionality. Diffusion directionality and flow are different. As shown in Fig. 4.2, if we drop ink into a media, the center of the ink moves when there is a flow. This motion can be described by one vector, which gives the directionality (orientation of the vector) and the flow speed (length of the vector). If we drop ink into a cup of water, water freely diffuses, and the shape becomes a sphere. If there is no flow, the center of the sphere does not move. This is random, incoherent motion of water molecules (see Fig. 1.3 in Chapter 1). It is called "isotropic" diffusion, in which water diffuses in all directions with the same amount. In this case, we need only one number, the diffusion constant (D), to describe the diffusion. The diffusion constant is related to the diameter of the sphere, and a sphere needs only one parameter to be uniquely determined. Things become more complicated when the shape of the ink becomes oval in a 2D plane or ellipsoid in 3D. This type of diffusion is called "anisotropic" diffusion, and the ellipsoid is called a "diffusion ellipsoid." This anisotropic diffusion is what often happens in biological tissues, namely, water tends to diffuse along

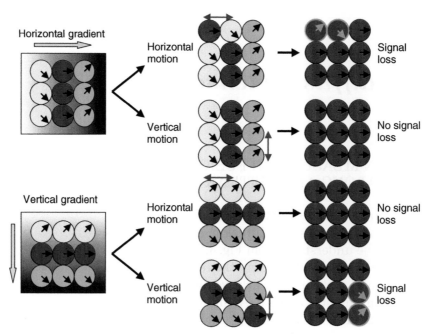

Fig. 4.1 Relationship between gradient orientation, molecular motion, and signal loss. When the horizontal gradient is used, molecular motions along the horizontal axis lead to signal loss.

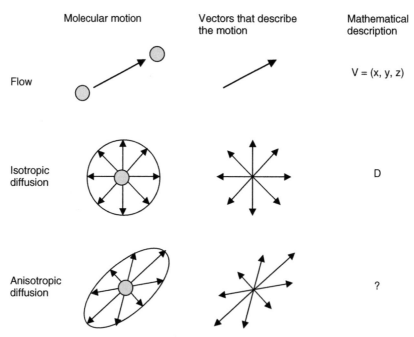

Fig. 4.2 Comparison between flow, isotropic, and anisotropic diffusion.

a preferential axis or axes. Apparently, we cannot describe this type of diffusion process using a single diffusion measurement or by a single diffusion constant.

This anisotropic diffusion is of great interest because it carries much information about the underlying anatomical architecture of living tissues (Fig. 4.3). Whenever there is ordered structures such as axonal tracts in nervous tissues or protein filaments in

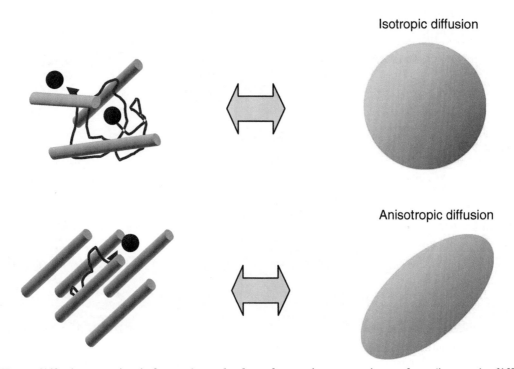

Fig. 4.3 Water diffusion carries information whether the environment is random (isotropic diffusion) or ordered (anisotropic diffusion).

muscle, water tends to diffuse along such structures. If we can determine the way water diffuses, we can obtain precious information about the object. This is exactly what we try to accomplish using diffusion tensor imaging (DTI).

Figure 4.4 shows the results of two diffusion constant measurements using the right–left and anterior–posterior gradient orientations. The results clearly show that the diffusion process inside the brain is highly anisotropic. It is of interest to us to characterize the way water diffuses inside the brain, from which we can extract information about axonal anatomy.

4.3   SIX PARAMETERS ARE NEEDED TO UNIQUELY DEFINE AN ELLIPSOID

To characterize the diffusion process with directionality (anisotropic diffusion), we need more elaborate diffusion measurement and data processing. One such method is called diffusion tensor imaging. Let us set aside diffusion measurement for a while and think how many parameters we need to define a circle, an oval, a sphere, and an ellipsoid (Fig. 4.5). A circle and a sphere need only one parameter, which is diameter, as described earlier. An oval needs three. We definitely need two numbers for the length of the longest and shortest axes, which describes the shape of the oval. The last one is needed to define the orientation of one of the axes.

The six parameters needed to define an ellipsoid are rather more complicated. We definitely need three lengths for the longest, shortest, and middle axes that are perpendicular to each other.

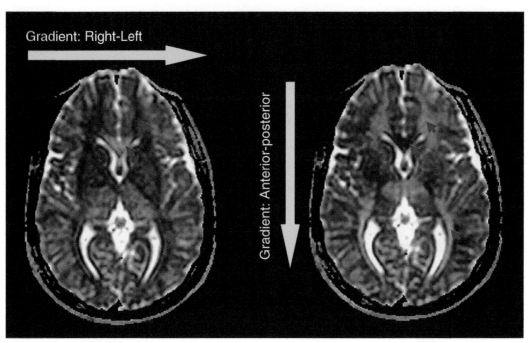

Fig. 4.4 Diffusion constant maps of a human brain using two different gradient orientations. Brain regions indicated by red arrows have markedly different diffusion constants between the two images.

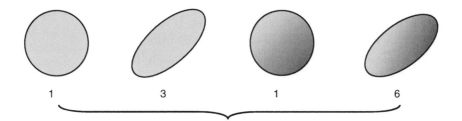

Fig. 4.5 The number of parameters needed to define an oval and an ellipsoid.

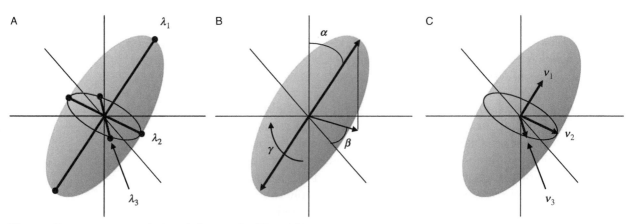

Fig. 4.6 Parameters needed to define a 3D ellipsoid.

These three lengths are usually called λ_1, λ_2, and λ_3, or "eigenvalues" (Fig. 4.6A). There are several ways to describe the orientation of the ellipsoid. Let us start with the orientation of the longest axis. If we use a polar-angle system, two angles (α and β) can define it. Unless the second and the third axes have the same length (cylindrical symmetry), a rotation angle about the longest axis is also needed to define the orientations of the shorter axes (angle γ in Fig. 4.6B). Another simpler way is to use three unit vectors to define the orientation of the principal axes (Fig. 4.6C). These three vectors are called $\mathbf{v_1}$, $\mathbf{v_2}$, and $\mathbf{v_3}$, or "eigenvectors." Because we need these six parameters to characterize an ellipsoid, six measurements of lengths along six arbitrary axes are needed to uniquely determine the ellipsoid.

4.4 DIFFUSION TENSOR IMAGING CHARACTERIZES THE DIFFUSION ELLIPSOID FROM MULTIPLE DIFFUSION CONSTANT MEASUREMENTS ALONG DIFFERENT DIRECTIONS

It is intuitive that anisotropic diffusion can be characterized by measuring the diffusion constant along many directions. Let us use a simple analogy to illustrate this point. Suppose we have a piece of paper that has tightly woven vertical fibers and loosely woven horizontal fibers (Fig. 4.7A). Let us cut the rectangular piece of paper into a circle and rotate it so that we no longer know the orientation of the vertical and horizontal axes (Fig. 4.7B). Here, our task is to decide the orientation of the vertical axes of the paper.

To answer this question, we can drop ink on the paper and see how the stain spreads (Fig. 4.7C and Fig 4.7D). We expect that the stain becomes oval instead of circular, and we can decide the orientation of the vertical fiber from the orientation of the longest axis. In this analogy, we can see the shape of the stain and find the orientation of the longest axis. However, what if the ink is invisible, and all we can do is to measure the length of the stain along arbitrary angles (Fig. 4.8A)? Can we define the shape of the stain by measuring the length along x and y axes (Fig. 4.8B)? As illustrated in Fig. 4.8B and Fig 4.8C, the two measurements are not enough to uniquely define the shape of the stain. To uniquely define the shape and orientation of a 2D oval, we need at least to know its length along three independent orientations (Fig. 4.8D). This is intuitively understandable if we think how many parameters we need to mathematically define an oval, which is three (Fig. 4.5). Of course, we can perform more measurements to obtain more information about the oval shape, which leads to an over-determined system. If there are no measurement errors, more than three measurements simply result in redundant data, but, in reality, we always have measurement errors, and the three-point measurement leads to significant inaccuracies for the estimation of the oval (Fig. 4.8E: no measurement error; Fig. 4.8F: with measurement error along an axis shown in red). If we have over-determined results, we can perform fitting to obtain a result that best satisfies all the measurement results (Fig. 4.8G: no measurement error; Fig. 4.8H: with measurement errors). This is very similar to the way we perform linear least-square fitting to estimate a simple

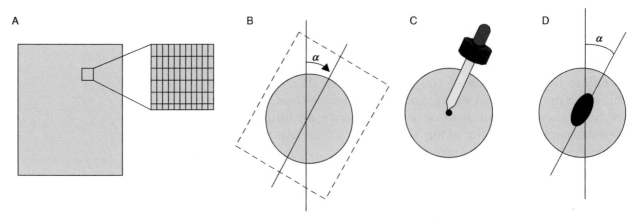

Fig. 4.7 Determination of fiber orientation of a piece of paper. A piece of paper has dense vertical fibers and sparse horizontal fibers (A). The paper is cut to a circle and rotated by an unknown angle, α (B). Ink is dropped on the paper (C), and the shape of the ink is observed (D). The angle of the longest axis tells the rotation angle, α.

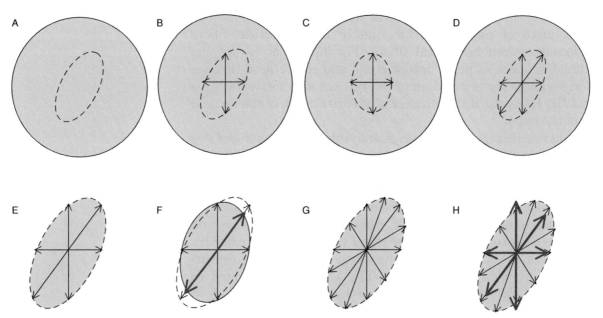

Fig. 4.8 Schematic for determination of an "invisible" oval ink stain. Suppose there is an ink stain that we cannot see but can measure its width along an arbitrary angle (A). Two measurements of its width cannot uniquely define the oval shape and orientation (B and C), but three measurements could (D). Note that the two measurement results of B and C are the same, although the stain shapes are different. If there is a measurement error, estimation of the ink shape become inaccurate (F: the red line indicates an erroneous measurement). If there are more than three measurements, the shape can be obtained using fitting (G), which is less sensitive to measurement errors (H).

linear equation; $y = \text{constant} + ax$. Mathematically, we need only two points to determine the parameters, a and constant. In order to increase accuracy, we usually obtain many more points and perform linear least-square fitting.

Because our brain is not 2D like the piece of paper, we have to extend this analogy to 3D space. The oval is now a 3D ellipsoid. We already know that we need six parameters to determine an ellipsoid. Naturally, we need six diffusion constants measured along six independent axes. Again, we can perform more than six measurements to more accurately define the ellipsoid under the existence of measurement errors. Our next task is to determine the six parameters (λ_1, λ_2, λ_3, $\mathbf{v}_1$, $\mathbf{v}_2$, and $\mathbf{v}_3$) from the six (or more) diffusion constant measurements. Here, we usually use a 3×3 tensor for the calculation and, thus, the imaging is called diffusion tensor imaging. The three parameters, λ_1, λ_2, and λ_3, are called eigenvalues and the three vectors, $\mathbf{v}_1$, $\mathbf{v}_2$, and $\mathbf{v}_3$, are called eigenvectors. The detail of the tensor calculation is provided in Chapter 5. At this point, therefore, let us assume that we can calculate the eigenvalues and eigenvectors from multiple diffusion constant measurements.

4.5   WATER MOLECULES PROBE MICROSCOPIC PROPERTIES OF THEIR ENVIRONMENT

As will be described later, the measurement time of water diffusion by NMR/MRI is approximately 10–100 ms. The diffusion constant of water inside the brain is typically 0.8–$0.9 \times 10^{-3}\,\mathrm{mm^2/s}$. Using Einstein's equation ($\sqrt{2Dt}$, where D is the diffusion constant and t is time), the average distance of the water diffusion during the measurement would be 4–15 μm. If there is any environmental property that confers water with anisotropy, it must have a smaller dimension than the diffusion distance. In neural structures, this could be protein filaments, cell membranes, cell organelles, and myelin sheaths. Their physical dimension is typically much smaller than 4–15 μm. One interesting fact is that diffusion anisotropy of *in vivo* and *ex vivo* brains is very similar, even though diffusion constant in the *ex vivo* samples is slower. Namely, if we perform an experiment of the ink stain inside a refrigerator, the size of the stain would become smaller due to a slower diffusion constant, but the shape would remain the same. This occurs because the diffusion distance is still sufficiently larger than the dimension of obstacles in the path of the diffusion. At $37°$, the diffusion constant of free water is approximately $3.0 \times 10^{-5}\,\mathrm{mm^2/s}$ and that of the parenchyma is less than $1/3$ ($0.8 \times 10^{-5}\,\mathrm{mm^2/s}$), suggesting that water molecules are bumping into many obstacles while they are diffusing and, thus, they are traveling long enough to probe structural properties of the environment. Strictly speaking, the equations to measure diffusion constants (Chapter 2) hold only when molecules are diffusing freely and thus diffusion process can be described by Gaussian distribution. If there are obstacles and boundaries, this assumption does not hold because the diffusion no longer obeys the Gaussian distribution (Eq. 3.2). The diffusion constant of the parenchyma could be truly slower than that of CSF because of higher viscosity, but majority of the reduction in the translational motion is likely due to the obstacles. Therefore, the diffusion constant calculated from Eq. 3.17 is called "apparent" diffusion constant (ADC).

4.6   HUMAN BRAIN WHITE MATTER HAS HIGH DIFFUSION ANISOTROPY

In the previous sections, we used a piece of paper to explain diffusion anisotropy, but of course, we are not interested in the anatomy of paper. Here, our interest is brain anatomy, especially the white matter. Measurement of diffusion anisotropy inside the white matter is of great interest because water tends to diffuse along axonal tracts and, thus, the anisotropy carries unique anatomical information of axonal architecture. There are three important facts that make DTI study of axonal anatomy especially exciting. First, the white matter looks homogenous in conventional MRI, and we cannot appreciate intra-white-matter anatomy very well. Second, study of axonal trajectory is not straightforward even in postmortem histology studies. The white matter in a freshly cut specimen is rather featureless. Most of our existing knowledge about human white matter anatomy has come from pathologic or accidental lesions of the brain, but we cannot control the size and location of the lesions in human studies. *In vivo* examination is of course impossible. Third, the DTI technique can capture the neuroanatomy of the entire brain. There are detailed studies of white matter anatomy of animals, done using invasive techniques such as injection of chemical tracer and subsequent histology analysis. However, the invasive approach can study only a small portion of neurons and cannot be used for global anatomical characterization. DTI is thus a technique that could provide us with anatomical information that no other methods can offer. It excels in the characterization of macroscopic white matter anatomy in a noninvasive manner.

We know from the previous section that the diffusion process reflects microscopic cellular architecture. Then why does DTI provide information about macroscopic white matter anatomy? This is a puzzling question. Diffusion anisotropy reflects microscopic anatomy (less than $10\,\mu m$), but our sampling resolution is coarse (typically 2 to 3 mm for each pixel). Inevitably, the microscopic information is averaged within a pixel. If there is a great deal of inhomogeneity of the microstructures within a pixel, the information from the pixel becomes featureless, i.e., diffusion looks isotropic. This leads to a "double-layered" structure in DTI information. The first layer is microscopic anatomy, which confers anisotropy to water diffusion. The second layer is macroscopic coherent arrangement of the anisotropic microscopic anatomy. Only when these two factors exist in a pixel can we observe diffusion anisotropy. This point will be discussed in more detail in later chapters, but it is important to keep this fact always in mind.

Chapter 5

Mathematics of diffusion tensor imaging

5.1   OUR TASK IS TO DETERMINE THE SIX PARAMETERS OF A DIFFUSION ELLIPSOID

Our task is to define the shape and orientation of the ellipsoid (called the diffusion ellipsoid). The most intuitive way is to measure diffusion constants along numerous orientations, from which the shape can be reconstructed. This type of direct measurement of diffusion anisotropy has actually become quite popular recently. An alternative way is to measure diffusion constants along a smaller number of orientations, from which the shape of the ellipsoid is calculated (note that here we are making the big assumption that diffusion property is elliptic). For this calculation, we need the aid of a mathematical procedure called "tensor" calculation and, thus, it is called diffusion tensor imaging (DTI).

First of all, I would like to make it clear that anisotropic diffusion (or the diffusion ellipsoid) cannot be characterized by measurements along three orthogonal axes, x, y, and z. This is for the same reason that two measurements cannot uniquely determine the shape of the oval, as explained in the previous chapter (Fig. 4.8). In order to fully characterize the diffusion ellipsoid, we need six parameters, as explained in Chapter 4. These are related to three numbers for the length of the longest (λ_1), middle (λ_2), and shortest (λ_3) axes, which define its shape, and three vectors ($\mathbf{v}_1$, $\mathbf{v}_2$, $\mathbf{v}_3$) to define the orientations of the axes (called principal axes). In order to keep track of these six parameters, we use a 3×3 tensor, called a diffusion tensor, $\overline{\overline{\mathbf{D}}}$, which is related to the six parameters by a process called "diagonalization."

$$\overline{\overline{\mathbf{D}}} = \begin{bmatrix} D_{xx} & D_{xy} & D_{xz} \\ D_{yx} & D_{yy} & D_{yz} \\ D_{zx} & D_{zy} & D_{zz} \end{bmatrix} \xrightarrow{\text{diagonalization}} \lambda_1, \lambda_2, \lambda_3, \mathbf{v}_1, \mathbf{v}_2, \mathbf{v}_3 \quad (5.1)$$

This diffusion tensor, $\overline{\overline{\mathbf{D}}}$, is a symmetric tensor, which means $D_{ij} = D_{ji}$, and, thus, there are six independent parameters, which makes sense because it intrinsically contains the six parameters of the diffusion ellipsoid.

Let us see how $\overline{\overline{\mathbf{D}}}$ and eigenvalues and eigenvectors are related in the following example (Fig. 5.1). Suppose we have a diffusion ellipsoid with a shape of $\lambda_1 = 2$, $\lambda_2 = 1$, and $\lambda_3 = 0.5$. When the longest

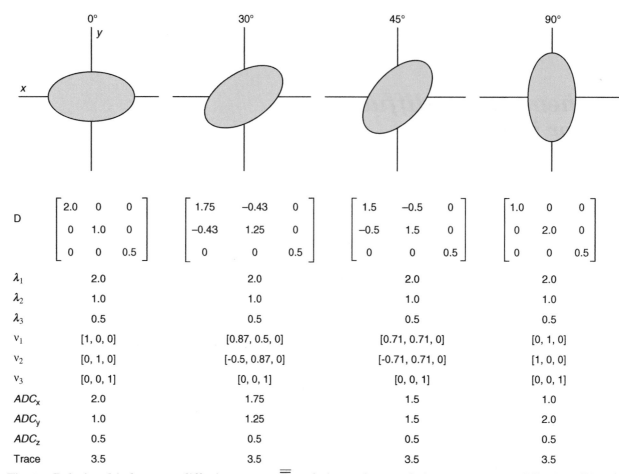

Fig. 5.1 Relationship between diffusion tensor, $\overline{\overline{D}}$ and eigenvalues and eigenvectors as a diffusion ellipsoid rotates about the z-axis (perpendicular to the plane). ADC_x, ADC_y, and ADC_z are apparent diffusion constants along the x, y, and z gradient axes (i.e., length of the ellipsoid along the x, y, and z axes). Trace is the sum of the diagonal elements, which is equal to the sum of three eigenvalues (λ_1, λ_2, λ_3), and is independent of rotation.

axis is aligned to the x-axis and the middle axis to the y-axis, the eigenvector $\mathbf{v}_1$, $\mathbf{v}_2$, and $\mathbf{v}_3$ coincide with the x, y, and z axes. The off-diagonal terms are 0 when the principal axes (eigenvectors) align to the x-y-z coordinate system. As the ellipsoid rotates about the z-axis, the eigenvectors and $\overline{\overline{D}}$ change. When they are not aligned to the coordinate system, we get, in this case, D_{xy}/D_{yx} off-diagonal terms, indicating that these terms carry information about the rotation about the z-axis. Similarly, D_{xz}/D_{zx} terms are related to rotation about the y-axis, and D_{yz}/D_{zy} terms are related to the x-axis. Note that eigenvalues are independent of the rotation because they carry information about the shape of the ellipsoid, which is independent of the rotation. The way the eigenvectors change is much more intuitive than the way $\overline{\overline{D}}$ does. Then why do we use $\overline{\overline{D}}$? In our experiments, we want to know the shape (eigenvalues) and orientation (eigenvectors) of the ellipsoid. Unfortunately, we cannot measure these six numbers directly. All we can do is to measure the extent of diffusion (equivalent of

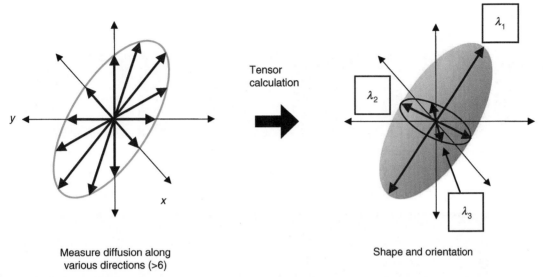

Fig. 5.2 A diffusion ellipsoid can be fully characterized from diffusion measurements along six independent axes.

lengths of the ellipsoid) along various orientations. The reason we use the tensor is that it is straightforward to relate these measurement results to the tensor elements. For example, the diffusion constant along the x-axis (ADC_x) is D_{xx} and along the y-axis is D_{yy}. In this way, the diagonal elements of the tensor are directly measurable. Note that D_{xy} is not the diffusion constant along the $45°$ axis from x and y-axes. If this is the case, the third example ($45°$ rotation) would have $D_{xy} = 2$. The off-diagonal elements carry information about the rotations but not the physical information about diffusion constants.

In order to determine these six elements of $\overline{\overline{D}}$, not surprisingly, we need to measure at least six diffusion constants along six independent axes (Fig. 5.2). In the following section, an actual experimental process to determine $\overline{\overline{D}}$ will be described.[1]

5.2   WE CAN OBTAIN THE SIX PARAMETERS FROM SEVEN DIFFUSION MEASUREMENTS

In order to measure six diffusion constants along six independent axes, at least seven diffusion-weighted images are needed. This is because at least two data points are needed to obtain a diffusion constant from a slope of signal attenuation. We commonly obtain a least-diffusion-weighted image, the intensity of which corresponds to S_0 in Eq. 5.2 (often called a $b0$ image, meaning $b = 0$). Then, by obtaining another diffusion-weighted image using, for example, x-gradient (S_x), we can calculate an apparent diffusion constant along x-axis (ADC_x). The same S_0 data can also be used to obtain ADC_y, together with a diffusion-weighted image with y-gradient (S_y). In this way, six diffusion constants can be obtained using various gradient combinations. Figure 5.3 shows an example of the

[1] Strictly speaking, the shape of the ellipsoid represents the iso-surface of mean displacement of water molecules (unit is mm), which is different from definition of diffusion constant (unit is mm^2/s). If we plot diffusion constants along multiple axes, the shape becomes more like a peanut.

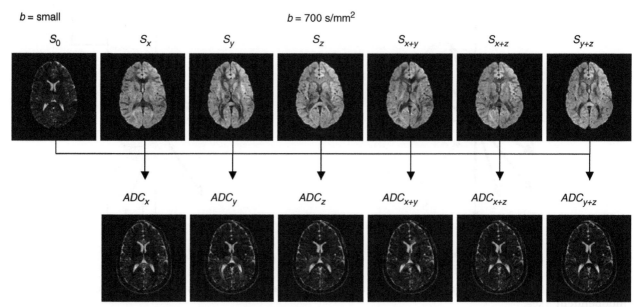

Fig. 5.3 At least one least-diffusion-weighted image and six diffusion-weighted images along six different axes are needed. In this figure, the least-diffusion-weighted image gives S_0 intensity and diffusion-weighted images along six independent gradient orientations give S_x, S_y, S_z, S_{x+y}, S_{x+z}, and S_{y+z} intensities. From these images, six diffusion constants can be obtained (ADC_x, ADC_y, ADC_z, ADC_{x+y}, ADC_{x+z}, and ADC_{y+z}). The least-diffusion-weighted images are often called $b = 0$ (or $b0$) or non-diffusion-weighted images, but it is experimentally impossible to obtain $b = 0$ images. Therefore, the convention of $b0$ images means, strictly speaking, "images with negligible diffusion weighting."

measurements in which diffusion-weighted images by x, y, z, $x + y$, $x + z$, $y + z$ gradients are used.

In the previous section, it was shown that signal attenuation by diffusion-weighting follows:

$$\frac{S}{S_0} = e^{-\gamma^2 G^2 \delta^2 (\Delta - \delta/3)D} = e^{-bD} \tag{5.2}$$

This equation is correct only for isotropic diffusion or for diffusion measurement along one axis. For a more complete expression for anisotropic media, we have to use:

$$\ln\left[\frac{S}{S_0}\right] = -\int_0^t \gamma^2 \left[\int_0^{t'} \overline{G(t'')}dt''\right] \bullet \overline{\overline{\mathbf{D}}} \bullet \left[\int_0^{t'} \overline{G(t'')}dt''\right] dt' \tag{5.3}$$

Again, if we solve this equation for the experiment with a pair of square-shaped gradients, we obtain:

$$\frac{S}{S_0} = e^{-\overline{\sqrt{b}}\overline{\overline{\mathbf{D}}}\,\overline{\sqrt{b}}^T} \tag{5.4}$$

where $\overline{\sqrt{b}}$ is $\gamma \overline{G}\delta\sqrt{(\Delta - \delta/3)}$. Here, $\overline{G}$ and $\overline{\sqrt{b}}$ are vectors because they contain information not only of gradient strength but also about the orientation.

In actual experiments, the six parameters in the $\overline{\overline{\mathbf{D}}}$ are what we want to determine and parameters, $\overline{G}$, δ, Δ, and γ are known parameters, and S_0 and S are the experimental results. Please note that this equation has a total of seven unknowns (six in $\overline{\overline{\mathbf{D}}}$ and S_0) and we have seven experimental results (S) with different $\overline{G}$ values. Therefore, it is solvable.

For example, if we use x-gradient ($\overline{G} = [G_x, 0, 0]$), obtain image intensity S_x, and plug this number into Eq. 5.4, the $\sqrt{b}\overline{\overline{\mathbf{D}}}\sqrt{b}$ portion becomes:

$$\gamma^2\delta^2(\Delta - \delta/3)(G_x, 0, 0)\begin{pmatrix} D_{xx} & D_{xy} & D_{xz} \\ D_{yx} & D_{yy} & D_{yz} \\ D_{zx} & D_{zy} & D_{zz} \end{pmatrix}\begin{pmatrix} G_x \\ 0 \\ 0 \end{pmatrix}$$

If we solve this part and put it back into Eq. 5.4, it becomes:

$$\frac{S_x}{S_0} = e^{-\gamma^2 G_x^2 \delta^2 (\Delta - \delta/3)D_{xx}} \tag{5.5}$$

from which we can obtain an element D_{xx}. Similarly, from experiments using the y- and z-gradient only, we can obtain D_{yy} and D_{zz}. If we apply the same strength of x and y-gradient simultaneously ($\overline{G} = [G_x, G_y, 0]$), we obtain the image intensity S_{xy}

$$\frac{S_{xy}}{S_0} = e^{-\gamma^2\delta^2(\Delta - \delta/3)(G_x^2 D_{xx} + 2G_x G_y D_{xy} + G_y^2 D_{yy})} \tag{5.6}$$

Because we already know D_{xx} and D_{yy}, we can calculate D_{xy} from this result. Similarly D_{xz} and D_{yz} can be obtained from experiments using $x + z$ gradient combination and $y + z$ gradient combination. In this way, from a total of seven diffusion-weighted images, the six elements of $\overline{\overline{\mathbf{D}}}$ can be determined. Then the diagonalization process gives us λ_1, λ_2, λ_3, $\mathbf{v}_1, \mathbf{v}_2$, and $\mathbf{v}_3$ of the diffusion ellipsoid. This calculation has to be repeated for each pixel, from which the diffusion tensor can be obtained for each pixel.

Extra attention should be paid to the fact that D_{xy} is not directly related to S_{xy} (Eq. 5.6) in the way that D_{xx} is related to S_x (Eq. 5.5); i.e., D_{xx} can be calculated directly from S_x, but this is not the case for D_{xy}.

5.3   DETERMINATION OF THE TENSOR ELEMENTS FROM A FITTING PROCESS

As described in the previous section, Eq. 5.4 has seven unknowns, which can be solved from seven measurements (or imaging) with seven different diffusion weightings (including one least-diffusion-weighted image). In practical situations, we often perform more than seven measurements using different gradient strengths, gradient orientations, or signal averaging (see Fig. 4.8). In this case, Eq. 5.4 is over-determined and the simple calculation process described here is not appropriate. Furthermore, the signal intensity with zero-diffusion weighting (S_0) sometimes cannot be obtained directly, because imaging techniques always use a finite amount of diffusion weighting. Therefore, we should use the fitting technique rather than solving it. For the fitting, we need to expand Eq. 5.4:

$$\ln\left[\frac{S}{S_0}\right] = -\sqrt{b}\overline{\overline{\mathbf{D}}}\sqrt{b}^T \rightarrow \ln(S) = \ln(S_0) - \sqrt{b}\overline{\overline{\mathbf{D}}}\sqrt{b}^T$$

$$\sqrt{b} = \left[\sqrt{b_x}, \sqrt{b_y}, \sqrt{b_z}\right] \tag{5.7}$$

$$b_{x,y,z} = \gamma^2 G_{x,y,z}^2 \delta^2(\Delta - \delta/3)$$

Next, we have to expand $\sqrt{b}\overline{\overline{\mathbf{D}}}\sqrt{b}^T$:

$$\sqrt{b}\overline{\overline{\mathbf{D}}}\sqrt{b}^T = \left[\sqrt{b_x}, \sqrt{b_y}, \sqrt{b_z}\right] \begin{bmatrix} D_{xx} & D_{xy} & D_{xz} \\ D_{yx} & D_{yy} & D_{yz} \\ D_{zx} & D_{zy} & D_{zz} \end{bmatrix} \begin{bmatrix} \sqrt{b_x} \\ \sqrt{b_y} \\ \sqrt{b_z} \end{bmatrix}$$

$$= D_{xx}b_x + D_{yy}b_y + D_{zz}b_z + 2D_{xy}\sqrt{b_x}\sqrt{b_y} + 2D_{xz}\sqrt{b_x}\sqrt{b_z}$$

$$+ 2D_{yz}\sqrt{b_y}\sqrt{b_z} = \overline{\mathbf{D}}\,\overline{\mathbf{b}} \tag{5.8}$$

Here, we have defined two new vectors, $\overline{\mathbf{D}}$ and $\overline{\mathbf{b}}$, which are defined as:

$$\overline{\mathbf{D}} = [D_{xx}, D_{yy}, D_{zz}, 2D_{xy}, 2D_{xz}, 2D_{yz}]$$

$$\overline{\mathbf{b}} = [b_x, b_y, b_z, \sqrt{b_x}\sqrt{b_y}, \sqrt{b_x}\sqrt{b_z}, \sqrt{b_y}\sqrt{b_z}] \tag{5.9}$$

By plugging this back into Eq. 5.7, it can be rewritten as:

$$\ln(S) = \ln(S_0) - \overline{\mathbf{D}}\overline{\mathbf{b}} \tag{5.10}$$

This equation is very similar to a simple linear equation, $y = \text{constant} - ax$, in which x is an independent variable (equivalent to the $\overline{\mathbf{b}}$ vector), y is the result (equivalent to image intensity, $\ln(S)$), constant is intercept to y-axis (equivalent to $\ln(S_0)$, which is the image intensity with $b = 0$), and a is the slope (equivalent to the $\overline{\mathbf{D}}$ vector). Similar to the equation, $y = \text{constant} - ax$, Eq. 5.10 can be solved by linear least-square fitting, although all variables are vectors, instead of scalar.

For example, in the experiment shown in Fig. 5.3, we used the gradient combination of $[0,0,0]$, $[1,0,0]$, $[0,1,0]$, $[0,0,1]$, $[1/\sqrt{2}, 1/\sqrt{2}, 0]$, $[1/\sqrt{2}, 0, 1/\sqrt{2}]$, $[0, 1/\sqrt{2}, 1/\sqrt{2}]$ (defined as $\overline{b_1}, \overline{b_2}, \ldots, \overline{b_7}$) and suppose we get image intensities $S_1, S_2, \ldots, S_7$

from each experiment; then, the entire experiment can be expressed as:

$$
\begin{bmatrix} S_1 \\ S_2 \\ S_3 \\ S_4 \\ S_5 \\ S_6 \\ S_7 \end{bmatrix} = \ln(S_0) - \overline{\mathbf{D}} \begin{bmatrix} \overline{\mathbf{b}}_1 \\ \overline{\mathbf{b}}_2 \\ \overline{\mathbf{b}}_3 \\ \overline{\mathbf{b}}_4 \\ \overline{\mathbf{b}}_5 \\ \overline{\mathbf{b}}_6 \\ \overline{\mathbf{b}}_7 \end{bmatrix}
$$

$$
= \ln(S_0) - \overline{\mathbf{D}} \begin{bmatrix} 0 & 0 & 0 & 0 & 0 & 0 \\ 1 & 0 & 0 & 0 & 0 & 0 \\ 0 & 1 & 0 & 0 & 0 & 0 \\ 0 & 0 & 1 & 0 & 0 & 0 \\ 1/2 & 1/2 & 0 & 1/2 & 0 & 0 \\ 1/2 & 0 & 1/2 & 0 & 1/2 & 0 \\ 0 & 1/2 & 1/2 & 0 & 0 & 1/2 \end{bmatrix} \tag{5.11}
$$

$$
\rightarrow \overline{S}^T = \ln(S_0) - \overline{\mathbf{D}}\overline{\overline{\mathbf{b}}}
$$

in which $\overline{\overline{\mathbf{b}}}$ is called the b-matrix. If 30 experiments with 30 different $\overline{\mathbf{b}}_i, (i = 1, 2, \dots, 30)$ are performed and 30 images $(S_1, \dots, S_{30})$ are obtained, these numbers can just be appended to the preceding equation, and the best estimated $\overline{\mathbf{D}}$ and S_0 can be obtained using multivariate linear fitting.

Chapter 6

Practical aspects of diffusion tensor imaging

6.1 TWO TYPES OF MOTION ARTIFACTS: GHOSTING AND COREGISTRATION ERROR

There are two reasons why we need to address and correct motion artifacts in DTI studies. One is a DTI-specific issue, and the other is a problem common to all quantitative MRI that requires multiple MR images such as T_2 map, magnetization transfer ratio, and perfusion map. The DTI-specific issue is degradation of diffusion-weighted images by ghosting effect (see Chapter 1, Fig. 1.3 and Fig. 6.1), as DTI is a technique hypersensitive to motion due to its use of strong diffusion gradients. Using a diffusion constant of normal brain tissue ($D = 0.8 \times 10^{-3}\,mm^2/s$) and typical gradient separation ($\Delta = 30\,ms$), we can calculate that water molecules move about $7\,\mu m$ on average during the measurement ($7\,\mu m = \sqrt{2D\Delta}$, based on Einstein's equation). Any bulk motions at or larger than this size can interfere with the diffusion measurements. In practice, it is almost impossible to remove all the bulk motions, no matter how securely the head is held inside a head coil, since even minute brain motions due to cardiac pulsation and respiratory motions can approach this amount. This small amount of bulk motion can lead to phase errors (see Fig. 1.10) and often result in severe ghosting effect.

This problem can be reduced by using single-shot techniques such as echo-planar imaging (EPI), which will be explained in the next section. However, it is still common that DWI is plagued by residual ghosting effect. Regions around the posterior fossa seem especially vulnerable to ghosting issues, which is likely to be related to brain pulsation. This problem could be ameliorated by using cardiac gating and avoiding a period when the pulsatile motion is at a maximum. However, this approach reduces the time efficiency of the scans and is affected by cardiac arrhythmia. Therefore, we need to carefully evaluate the cost–benefit function of cardiac gating. For example, if one is interested in diagnosis of major white matter malformation, a small amount of ghosting in some out of many DWIs may not be an issue. If one is quantifying diffusion anisotropy to detect 5% difference between normal controls and patients, cardiac gating would be the right choice. An effective and practical approach could be to inspect each DWI to find apparent ghosting. If images are corrupted, they should be removed from subsequent tensor calculation.

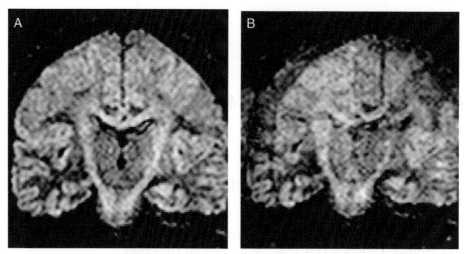

Fig. 6.1 Examples of diffusion-weighted images without (A) and with (B) ghosting.

In theory, we only need six DWIs to calculate a tensor field. However, in order to enhance the signal-to-noise ratio (SNR), we usually record more than 30 DWIs. As such, we can usually afford to remove several corrupted DWIs without sacrificing the SNR significantly. For the same reason, it is not recommended to carry out signal averaging "on-the-fly" by the scanner. Instead, repeated measurements should be made and inspected individually; once corrupted images are averaged with noncorrupted ones, it becomes more difficult to determine the source of the ghosting effect. The disadvantages of this visual inspection approach are that it is time consuming, and the detection of ghosting effect relies on subjective judgment. An automated approach for ghost detection would be desirable for future research.

The second motion-related problem exists in image coregistration (see Chapter 1, Fig. 1.3). DTI typically takes 5–15 min of scanning time during which 30–90 DWIs are acquired. If the subject moves more than the size of pixels (2–3 mm) during the scan, pixels in the DWIs are not coregistered. This registration problem is very different from the image corruption problem discussed earlier. Even though they are both caused by subject motion, they have different ramifications. Once image corruption occurs, the only measure we can take is to remove it from data processing. However, the misregistration can theoretically be corrected by image alignment via post-processing. Typically, a 3D image alignment of multiple images from the same subjects is carried out. This can be achieved by six-mode rigid-body rotation and translation. Free software is available for this purpose. Note that the realignment process involves interpolation of the pixels, which inevitably leads to a smoothing effect (loss of resolution and SNR enhancement). This remedy does not save data with severe motion problems because such data are often plagued by extensive ghosting, and there are an increasing number of problematic 3D images in which different slices are imaged at different head positions.

Coregistration tools cannot easily fix misalignment of slices within 3D volume data.

6.2 WE USE ECHO-PLANAR IMAGING TO PERFORM DIFFUSION TENSOR IMAGING

As explained in Chapter 1, Section 1.3.3, coherent motions lead to phase shift of the signal. It is important to understand the effect of this motion-related phase shift. For example, if one wants to obtain an image of a 128×128 matrix, the scanner acquires raw data (called time-domain data or k-space data) with a 128×128 resolution. In conventional imaging, the k-space is recorded line-by-line, requiring 128 independent scans, with each scan corresponding to one line. In the k-space, both phase and magnitude information are recorded, which can be converted to spatial location and intensity (i.e., image) after Fourier transformation (FT). Once nonreproducible phase shift is introduced at each scan, this leads to misregistration of proton signals after the FT, appeared as "ghosting" (Fig. 6.1B).

The most common and effective way for solving this problem is to use techniques such as single-shot echo-planar imaging (SS-EPI), in which the entire k-space is recorded within one scan. In this case, even if the diffusion weighting and bulk motions cause the phase shift, the entire k-space obtains the same amount of phase error, which, in theory, does not have any effect after FT. Unless there is a severe motion problem, the SS-EPI can address most of the motion-related ghosting issues. On the other hand, SS-EPI has its own problems. First, imaging resolution is limited. The length of the echo train[1] can usually only go up to 128 because there is not much signal left after the long echo train. If a field of view (FOV) of 240 mm is used, the resolution is up to 1.875 mm. Because there is not much signal left after 128-echo acquisition, a longer echo train, such as 144 echoes, does not lead to real resolution enhancement. Rather, the longer train leads to longer echo time and lower signal to noise. The second issue is image distortion, which will be discussed in more detail later in this chapter. To ameliorate these SS-EPI-related issues, other types of data acquisition have been postulated, such as segmented EPI, segmented spiral scans, and PROPELLAR acquisition. These approaches need some kind of "navigation" system to monitor and correct the phase shift via post-processing. Although these sophisticated approaches could be important for future applications, SS-EPI remains widely used, due especially to recent advances in parallel imaging technology that can dramatically reduce these problems.

There are several points that are worth stressing. First, the phase errors caused by subject motion always exist in conventional MRI. However, its extent is far larger in the diffusion-weighted images due to the application of a pair of large diffusion-weighting gradients. Second, it is very important to understand that SS-EPI is robust against motion-related artifacts not simply because it is rapid, but because it is less sensitive to phase errors. Third,

[1] The acquired signal is often called "echo." When signal equivalent of 128 scans are acquired in EPI, a train of 128 echo signals are recorded, which is called "echo train."

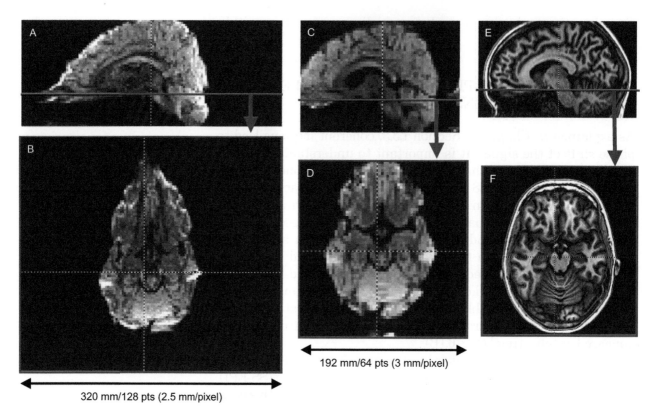

320 mm/128 pts (2.5 mm/pixel)

192 mm/64 pts (3 mm/pixel)

Fig. 6.2 The relationship between the length of echo train, resolution, and image distortion. Upper row shows mid-sagittal images and red lines indicate locations of the axial slices in the bottom row. Images A and B are obtained using a 128-echo train with 320-mm field of view (2.5 mm pixel resolution). Images C and D are acquired with a 64-echo train. To maximize the pixel resolution, the field of view is minimized to 192 mm (3 mm pixel resolution). Note that the field of view is too small, and a part of the frontal lobe is folded into the back of the head. Image distortion is much reduced with the 64-echo scan. Images E and F are T_1-weighted images as anatomical guidance.

the motion artifacts discussed earlier should not be confused with coregistration errors among the diffusion-weighted images. The ADC map or diffusion tensor imaging always requires multiple diffusion-weighted images from which many parameters are calculated on a pixel-by-pixel basis. Therefore, the location of each pixel has to be exactly coregistered.

As mentioned earlier, two most notable shortcomings of single-shot EPI is the limitation in resolution and image distortion due to B_0 susceptibility problems (Fig. 6.2). These two issues are related. The single-shot EPI acquires all phase-encoded lines after one excitation. As the echo train length gets longer, the extent of image distortion increases; i.e., the higher the resolution, the more distortion we get.

This is an inherent limitation of EPI. The distortion is caused by magnetic field (B_0) inhomogeneity, which is especially severe at the boundary of air and tissue; most notably around the sinus, such as inferior regions of the frontal lobe, the anterior pole of the temporal lobe, and the pons. The amount of distortion increases as the field strength increases. A 3T scanner produces more distortion than a 1.5T scanner. There are several ways to minimize this problem.

First, we can use a lower resolution as much as possible. Second, there are techniques that measure the B_0 field distortion, calculate a map of image distortion, and undo the distortion. This approach requires an additional scan for the field mapping. Therefore, it cannot be applied to existing data unless such a scan has been acquired for each study. This approach should be able to correct low-frequency distortion (global distortion) fairly accurately, but it is a challenging task to correct high-frequency distortion (severe local distortion). One of the most effective approaches in recent technologies is probably parallel imaging, which will be discussed in more detail in Section 6.5.

6.3   THE AMOUNT OF DIFFUSION-WEIGHTING IS CONSTRAINED BY THE ECHO TIME

The diffusion constant in healthy brain tissue is about 0.8–$1.0 \times 10^{-3}\,\text{mm}^2/\text{s}$. If $b = 1000\,\text{s/mm}^2$ is used, it would lead to loss of signal by approximately half:

$$e^{-1000 \times 0.8 \times 10^{-3}} = 0.45$$

In order to maximize the efficiency of diffusion weighting, the entire echo time is usually filled with diffusion-weighting gradients. This means gradient length (δ) is set close to the gradient separation (Δ) (Fig. 6.3A). Let us set $\Delta = \delta + 5\,\text{ms}$ to accommodate a $180°$ RF pulse. Typical clinical scanners are equipped with a gradient system that can be driven up to 20 or $40\,\text{mT/m}$. To achieve $b = 1000\,\text{mm}^2/\text{s}$, we need δ of approximately $35\,\text{ms}$ for $20\,\text{mT/m}$ systems and $21\,\text{ms}$ for $40\,\text{mT/m}$ (remember, $b = \gamma^2 G^2 \delta^2 (\Delta - \delta/3)$, where $\gamma = 2.675 \times 10^8/\text{s, T}$). The length of echo time (TE) must then be approximately 50–$70\,\text{ms}$ to accommodate the diffusion-weighting gradient pulses (Fig. 6.3A). In reality, we usually use echo-planar imaging. The length of the echo train is typically 30–$50\,\text{ms}$, depending of the bandwidth, slew rate, and imaging matrix size. To accommodate this long echo train, the echo time needs to be lengthened by the length of the echo train, as shown in Fig. 6.3B (red arrows). By combining the time required for diffusion weighting and the echo train, the echo time needs to be 80–$120\,\text{ms}$. To shorten this long echo time, it is common to use an asymmetrical echo train, as shown in Fig. 6.3C. For example, by using 25% truncation (75% k-space sampling), echo time can be shortened by half of the echo train length (15–$25\,\text{ms}$).

6.4   THERE ARE VARIOUS k-SPACE SAMPLING SCHEMES

To achieve higher SNR (shorter echo time) and less image distortion, the ideal approach is to minimize the length of the echo train length in SS-EPI. There are several strategies to achieve this goal.

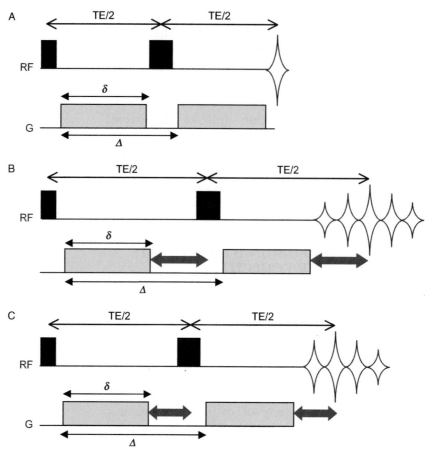

Fig. 6.3 Relationship between echo times and diffusion-weighting parameters. For a simple spin echo (A), the echo time can be filled with diffusion-weighting gradients. For an echo-planar sequence (B), the echo time must be increased to accommodate the echo train. The amount of extra echo time required is indicated by red arrows (half the length of the echo train). To shorten the echo time, an asymmetrical echo train is often used (C). Although only 4–5 echoes are drawn in this schematic diagram, the actual echo train contains 96–128 echoes, depending on the number of phase-encoding steps (the number of pixels in the image).

In terms of hardware, it can be achieved by stronger gradient, higher-gradient slew rates, and wider bandwidth. Partial k-space coverage is also an option. In terms of imaging parameters, minimizing the imaging matrix is the most effective way to shorten the echo train length. The matrix size, field of view (FOV), and spatial resolution has the following relationship; FOV/matrix size = resolution. If one does not want to sacrifice the spatial resolution, a reduction of matrix size needs to be accompanied by a reduction in FOV. For example, if 2 mm of resolution is needed, FOV/matrix size could be 256×256 mm/128×128 or 192×192 mm/96×96. One of the most commonly used image resolution for DTI is 2.5 mm. For this resolution, 240×240 mm/96×96 or 200×200 mm/80×80 can be used, depending on the brain size. If the research involves only pediatric brains, it makes sense to employ smaller FOV and matrix sizes. One drawback of adjusting FOV for each subject is that image quality would not be consistent among subjects. Larger FOV and imaging matrix lead to less SNR and more distortion.

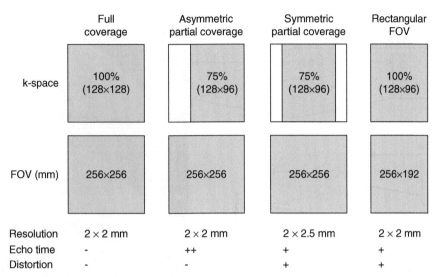

Fig. 6.4 Various types of k-space sampling schemes and FOVs. The k-space, FOV, and resolution are indicated by frequency × phase-encoding steps. For echo time and distortion, −, +, and ++ indicate the extent of improvement, while − indicates the same as full coverage.

An alternate approach is to reduce the number of phase-encoding steps, which leads to partial k-space coverage, as shown in Fig. 6.4. There are two types of partial coverage schemes: symmetric and asymmetric. Symmetric partial coverage leads to a shorter echo time (not as efficient as asymmetric coverage) and less image distortion. The tradeoff is a reduction in spatial resolution in the phase encoding direction. Asymmetric k-space coverage can efficiently shorten the echo time, as explained in the previous section. In this approach, the degradation in spatial resolution is not as much as that of symmetric coverage, because an edge of k-space is sampled (the edges of a k-space carry information about high spatial resolution). A shorter echo train allows shorter TE and, thus, higher SNR, but the extent of image distortion is similar to that of full coverage. Figures 6.5A and 6.5B show a comparison of full and asymmetric partial k-space coverage using actual images. The improvement in SNR is obvious due to a much shorter echo time, while image distortion is approximately the same and a small amount of blurring is evident. The asymmetric k-space sampling is beneficial and widely used for DTI studies. On the other hand, our experience suggests that the improvement diminishes when a parallel imaging method is used, as will be discussed in the next section.

A rectangular FOV is also an effective approach to shorten the echo train length. This approach can be used only for elongated objects. Because brains have oval shapes in axial slices, it is a good idea, in theory, to use a rectangular FOV. However, there are several complications. First, let us assume that FOV is 256 (anterior–posterior) × 192 (right–left) mm and imaging matrix is 128 × 96 to obtain 2.0-mm spatial resolution. Our first option is to set the frequency encoding along the right–left orientation and phase encoding along anterior–posterior

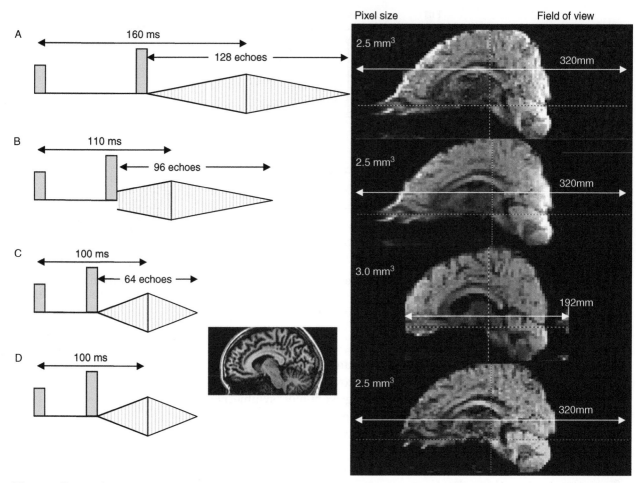

Fig. 6.5 Examples of SS-EPI images with various imaging parameters. The small inset image is a T_1-weighted image as a reference. Full k-space coverage with a 128×128 matrix (2.5 mm resolution) requires 160 ms of echo time and has severe image distortion and low SNR (A). Asymmetrical partial k-space coverage (75%) leads to a shorter echo time (110 ms) and higher SNR. However, it still suffers from severe distortion (B). Reduction in matrix size to 64×64 drastically reduces image distortion, but the penalty is low spatial resolution (C). With 3-mm resolution, FOV is 192 mm (= 64 points $\times$ 3 mm), which is not large enough for an average adult brain (part of the brain is folded in this image). Parallel imaging (SENSE) can effectively address these issues by providing a shorter echo time (100 ms) and less distortion (D).

(phase encoding for the long axis). The second option is to set the frequency encoding along the anterior–posterior orientation and phase encoding is right–left (phase encoding for the short axis). This is a feasible option to get shorter echo times and less distortion. However, the way the image distorts with respect to anatomy is an issue. In SS-EPI, the susceptibility distortion occurs mostly along the phase encoding orientation (Fig. 6.6). In the second option, the brain distorts along the right–left axis (Fig. 6.6A). Distortion is more benign when it occurs along the anterior–posterior axis because it does not affect the intrinsic right–left symmetry of the brain geometry. This is important for daily radiological diagnosis, in which the symmetry is often important information. The symmetry may also be beneficial for computational neuroanatomy because some important anatomical features such as brain midline are well preserved.

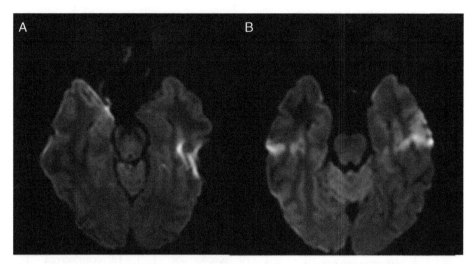

Fig. 6.6 Examples of images with phase encoding in the right–left orientation (A) and the anterior–posterior orientation (B). For the first example, a rectangle (240 × 192 mm) is used for FOV, which can reduce the phase encoding steps (96 readout points *versus* 72 phase encoding points) and, thus, the length of echo train is shortened. However, image distortion occurs along the right–left axis, which affects the right-left symmetry of the brain. In the second example, the phase encoding is along the anterior–posterior direction with a square FOV (240 mm × 240 mm FOV and 96 × 96 matrix). This leads to more phase encoding steps and, thus, longer echo train and more distortion. However, the distortion in the anterior–posterior orientation does not disturb the symmetry of the brain and is more benign.

Considering all these factors, square FOV with the phase encoding steps along the anterior–posterior orientation is still the parameter of choice for many published studies.

In conclusion, all the approaches for k-space and FOV setup have pros and cons in terms of echo time, spatial resolution, and image distortion. It is difficult to definitively conclude which approach is the best. For conventional SS-EPI, the asymmetric partial k-space coverage is beneficial, but probably the best approach is to use parallel imaging that can fundamentally change the relationship between k-space coverage, FOV, spatial resolution, echo time, and distortion issues, as discussed in the following section.

6.5   PARALLEL IMAGING IS GOOD NEWS FOR DTI

There are several implementations of parallel imaging techniques, and they are known by different acronyms, such as SENSE, GRAPPA, SMASH, and ASSET, by different scanner manufacturers. These various implementations have differences in technical details, yet they all achieve the same effect, i.e., shorter length of the echo train. For example, using conventional imaging hardware, if one wants 128-pixel resolution across the phase-encoding direction, the single-shot imaging needs to acquire 128 echoes at once. However, using parallel imaging with a typical reduction factor of two, the echo train length can be shortened to 64 without sacrificing the pixel resolution (128 pixels). This can effectively reduce the amount of distortion and shorten the echo time (Fig. 6.5D). Without post-processing, this approach can reduce the distortion from its

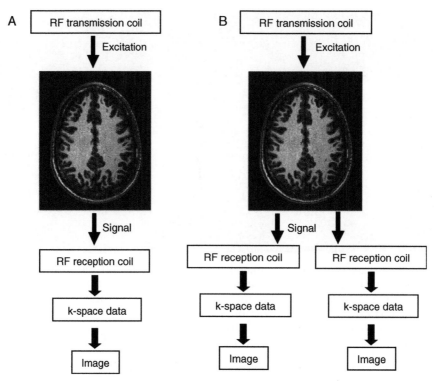

Fig. 6.7 Schematic diagrams of conventional imaging (A) and parallel imaging (B). For parallel imaging, a two-channel system with two-coil reception is used as an example.

source. This approach needs multiple channels and a special head coil to simultaneously and independently acquire MR signals. Therefore, parallel imaging is possible only when one has access to such equipment.

A simple diagram of parallel imaging is shown in Fig. 6.7. In conventional imaging (Fig. 6.7A), there is only one receiver channel. Signals from a reception coil (or coils) produce one k-space frame and subsequently one image. In parallel imaging, more than two receiver channels are used, which are attached to independent receiver coils, each producing a k-space frame and an image independently. Thus, if we have eight RF reception coils connected to an eight-channel system, we would get eight images simultaneously.

The reason why parallel imaging can shorten the echo train length is that we can reduce the sampling density of each channel. However, this results in a smaller FOV than the object (Fig. 6.8). As such, this approach is not feasible for conventional imaging because it generates a "folding" problem. With a multichannel system, however, we can "unfold" this image using multiple images (Fig. 6.9). In this way, we can use an FOV smaller than the object and shorten the echo train length. An example of the improvement can be seen in Fig. 6.5D. Although this approach cannot completely remove image distortion, it can drastically reduce it. One drawback of this approach is the loss of SNR, which can, however, be offset by the shorter echo time. The number of

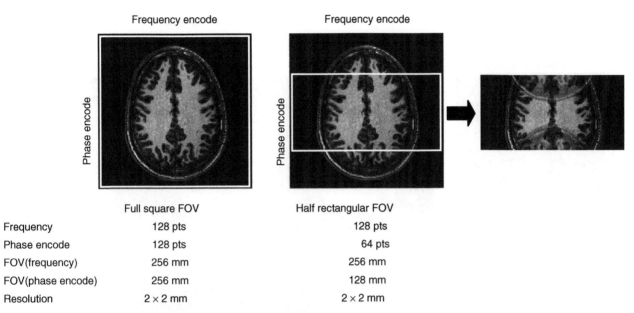

	Full square FOV	Half rectangular FOV
Frequency	128 pts	128 pts
Phase encode	128 pts	64 pts
FOV(frequency)	256 mm	256 mm
FOV(phase encode)	256 mm	128 mm
Resolution	2×2 mm	2×2 mm

Fig. 6.8 Parallel imaging allows us to reduce FOV in the phase encode orientation. In this example, a FOV (and the number of phase encode steps) is reduced by a factor of two, which makes the FOV smaller than the brain. This half-FOV approach leads to image folding, which is not usable for conventional imaging.

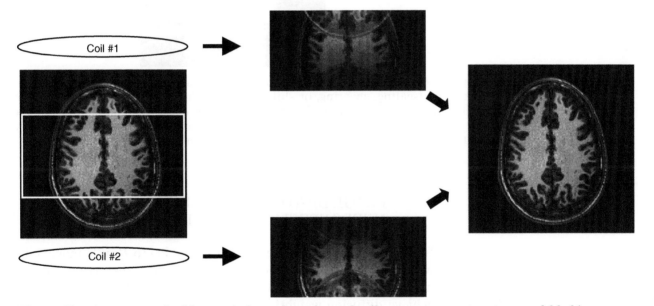

Fig. 6.9 Two images acquired by two independent channels allow us to reconstruct one unfolded image.

the reduction factor (called "p" factor) can be larger than two. The higher the p factor, the smaller the FOV and the shorter the echo train length. However, SNR and signal homogeneity degrades as the p factor increases because of difficulties in image reconstruction and unfolding. Currently, p factors of 2–3 are realistic numbers. We expect that most DTI studies will be based on this technique in the near future.

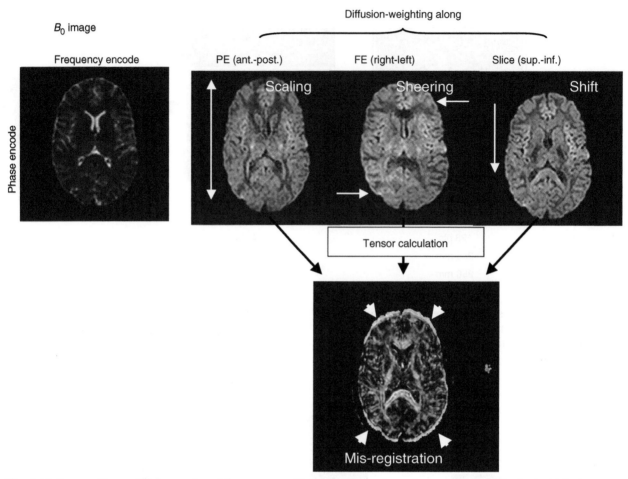

Fig. 6.10 Image distortion issues by eddy currents. Depending on the orientation of diffusion-weighting gradients, different types of distortion (scaling, sheering, or shift) occur, as shown in the figure. For tensor calculation, these images with different distortion lead to misregistration of pixels. This manifests as a high-anisotropy ring around the brain. PE, FE, and slice represent phase-encoding (anterior–posterior), frequency-encoding (right–left), and slice excitation (superior–inferior) orientations.

6.6   IMAGE DISTORTION BY EDDY CURRENT NEEDS SPECIAL ATTENTION

In the previous section, image distortion due to B_0 inhomogeneity was discussed. There is another major source of distortion that we need to be aware of, which is the diffusion weighting gradient. To achieve diffusion weighting, we need to apply a pair of gradients to sensitize the images to the diffusion process. To apply sufficient diffusion weighting (enough b-value) within the shortest amount of time (we want to keep echo time the shortest to ensure high signal intensity and SNR), we usually use the maximum gradient strength available. This type of strong gradient application is not often used for other types of MR imaging; however, it is quite common in DTI scans. As a consequence, we often see artifacts due to instrumental imperfections, which may not occur in other types of MR images (Fig. 6.10). One of the most common problems is caused by the so-called eddy current effect, which is due to a finite amount of residual gradient fields that linger after the diffusion-weighting gradients are turned off. If this lingering gradient overlaps with signal detection, unwanted image distortion occurs.

It is very important to understand that this image distortion by eddy current (eddy current distortion) has very different consequences from the EPI-related image distortion by B_0 susceptibility (B_0 distortion). The B_0 distortion affects all $b0$ and diffusion-weighted images in the same way. The eddy current distortion affects each diffusion-weighted image in different ways. This means that pixels in different DWIs are not coregistered, and the tensor calculation is no longer accurate. This causes erroneously high diffusion anisotropy for pixels at tissue boundaries such as brain parenchyma and CSF, because these pixels are sometimes inside the brain and sometimes outside, depending on the gradient direction. In most cases, the eddy-current-induced distortions are mostly linear and global, which can be corrected in post-processing. For example, the $b0$ images can be used as nondistorted reference images to unwrap the distorted DWIs. In doing so, the 12-mode linear transformation (affine transformation) is highly effective, and the relevant freeware is widely available.

Although the post-processing correction is useful, it is also important to minimize the eddy current itself. For example, pulse sequences can be designed to reduce it effectively. One of the most widely used approaches is the so-called double-echo sequence, which is shown in Chapter 2, Fig. 2.2C. The lingering gradient can be canceled out by the consecutive applications of positive and negative gradients. The most important measure is to make sure that gradient systems are tuned appropriately. If properly set up, modern gradient systems are not supposed to have severe eddy current problems. Routine assessment of eddy current distortion and proper maintenance are important for this issue.

6.7 DTI RESULTS MAY DIFFER IF SPATIAL RESOLUTION AND SNR ARE NOT THE SAME

In previous sections, pulse sequences and k-space sampling schemes (including FOV and imaging matrix) are discussed. When we design DTI studies, there are many more parameters that need to be set up. These include imaging time, imaging resolution, echo time (TE), repetition time (TR), gradient orientations, b-value, and the number of $b0$ images. Among these parameters, we will discuss spatial resolution and SNR in more detail.

It is a common feature of MR studies that different biological events have the same impact on a particular aspect of the MR signal; for example, both decrease in myelin concentration and vasogenic edema result in an increase in T_2 relaxation time. This kind of degeneration of biological information is to some degree inevitable in MR studies, as long as we are observing the physical and chemical properties of water molecules, which are indirect indicators of the underlying biological events. Diffusion anisotropy is especially a complex contrast, because it is sensitive not only to the physical and chemical environment of water molecules (microscopic anatomy such as myelin and axonal density), but also to the homogeneity of fiber orientation within a pixel, which is a far more macroscopic biological property (as large as the pixel size). As a result, diffusion anisotropy can be very sensitive to spatial resolution. In fact, the cortex has low anisotropy not because it does not contain myelin or axons, but because it has a complicated fiber organization with respect to the pixel dimension. If we have infinitesimal pixels, the gray matter could have high anisotropy too. We can see that boundaries of two major white matter tracts always have low anisotropy. This is again due to the mixture of fibers with different orientations. In T_1- and T_2-weighted images, the white matter has rather homogeneous T_1/T_2 values, and thus differences in spatial resolution do not have a large impact on the results. However, the white matter architecture is highly convoluted, and anisotropy values could be sensitive to image resolution; i.e., the lower the spatial resolution, the lower the diffusion anisotropy. This makes it rather complicated to compare DTI results across different brain sizes (e.g., brain development studies). It is important to note that absolute spatial resolution in geometric units (e.g., millimeters) is different from anatomical resolution. If we use the same pixel size for brains with different sizes, there are fewer pixels in the smaller brain (the same spatial resolution but different anatomical resolution). Strictly speaking, this may lead to lower diffusion anisotropy due to more mixture of fibers within a pixel. In order to avoid this, we need to have the same amount of pixels within the brains (the same anatomical resolution). This can be achieved by dynamically changing FOV based on the brain size. However, this leads to a different issue: impact of signal-to-noise ratio (SNR) on anisotropy measurements.

SNR is related to spatial resolution and imaging time. We all strive for the highest possible SNR but imaging time is often limited in clinical studies and, therefore, we have to accept a certain amount of noise in our measurements. It is commonly accepted that higher noise leads to increase in standard deviations. We have to be very cautious because in addition to larger standard deviations, noise is known to introduce bias in anisotropy calculations. It has been reported that anisotropy tends to increase as the SNR goes down. Thus, if we have two study groups, and data in the two groups have different SNRs, there is a chance that significant difference in diffusion anisotropy could exist simply due to the SNR difference. Therefore, it is important to understand these properties of DTI measurements and possible biases due to differences in anatomical resolution and SNR, so that we can carefully and correctly interpret DTI results.

6.8   SELECTION OF b-MATRIX

In Chapter 5, the gradient orientations in the b-matrix were some-what arbitrarily determined using the combination $[0,0,0]$, $[1,0,0]$, $[0,1,0]$, $[0,0,1]$, $[1/\sqrt{2}, 1/\sqrt{2}, 0]$, $[1/\sqrt{2}, 0, 1/\sqrt{2}]$, $[0, 1/\sqrt{2}, 1/\sqrt{2}]$. In reality, we should use the optimal b-matrix. However, it can be com-plicated as we need to consider the absolute strength of b-value, gradient orientations, and the total number of gradient orienta-tions. If we assume that we have 2 min to obtain DTI data, during which we can acquire approximately 20 DWIs, then we have several choices for imaging protocols. For example, we can choose a scheme with $b = 1000\,\text{s/mm}^2$ and six gradient orientations (a total of seven diffusion-weighted images, including one image at close to $b = 0\,\text{s/mm}^2$). To enhance the signal-to-noise ratio (SNR), the experiments can be repeated three times, giving rise to a total of 21 diffusion-weighted images. Alternatively, the seven diffusion-weighted images can be acquired with three different b-values (e.g., $b = 500$, 1000, and $1500\,\text{s/mm}^2$). Finally, diffusion-weighted images with 20 different gradient orientations with $b = 1000\,\text{s/mm}^2$ and one repeti-tion (no signal averaging) can be acquired within the similar amount of time.

6.8.1   Strength of the b-value

If the b-value is too small, the signal loss by diffusion process would become too small, and we cannot determine the signal decay accurately. If the b-value is too large, we would observe large signal decay, and signal intensity may reach the noise level. Obviously, there is an optimal range for the b-value, which depends on diffusion constants of the sample and SNR. This is not an easy issue because the range of apparent diffusion constants in the brain can be substantial due to large anisotropy (typically $0.1–1.6 \times 10^{-3}\,\text{mm}^2/\text{s}$). Practically, $b = 600–1200\,\text{s/mm}^2$ is the range most often used for clinical studies. As mentioned earlier, the average diffusion constant (the trace of tensor) of water inside the brain is typically $0.8–0.9 \times 10^{-3}\,\text{mm}^2/\text{s}$. When $b = 1000\,\text{s/mm}^2$ is used, signal attenuation would be:

$$e^{-1000 \times 0.8 \times 10^{-3}} = 0.45$$

In highly anisotropic regions, when diffusion is measured perpendicular to the fiber, it is not uncommon that apparent diffu-sion becomes very small. For example, if FA = 0.9 (see Section 7.1 for the definition of FA) and diffusion weighting is applied to perpendi-cular to the fiber orientation, the apparent diffusion constant could be as small as $0.1 \times 10^{-3}\,\text{mm}^2/\text{s}$, and signal attenuation is only 10%. When the measurement is aligned to the fiber orientation, the apparent diffusion constant can be more than $1.6 \times 10^{-3}\,\text{mm}^2/\text{s}$. Then the signal attenuation becomes 80%. If we use $b = 300\,\text{s/mm}^2$,

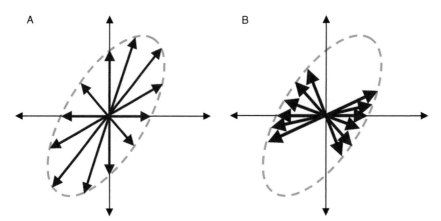

Fig. 6.11 Even (A) and skewed (B) distribution of gradient orientations to infer the shape of a diffusion ellipsoid. The former approach gives better-fitting power to estimate the ellipsoid.

the signal attenuation range becomes 3–32%, and $b = 3000\,\text{s/mm}^2$ leads to 26–99%.

The use of different b-values seems a logical step. With higher b-values, we can expect more signal decay for low-diffusion regions (fiber orientation is perpendicular to the gradient orientation), and with smaller b-values, less signal decay for high-diffusion regions (fiber orientation is parallel to the gradient orientation). However, scanning time would increase proportionally to the number of b-values. In addition, since the echo time is decided by the largest b-value, the advantage of using smaller b-values cannot be fully taken due to the long echo time.

6.8.2   Orientation of applied gradients

For measurement orientation (gradient combination), even sampling of the 3D space is the optimal method. This is logical, because our task is to define the shape and orientation of the diffusion ellipsoid, which we do not know *a priori* (Fig. 6.11). The gradient combinations we used in Fig. 5.3 of Chapter 5 do not distribute uniformly in the 3D space and, thus, it is not an optimum sampling scheme. Unfortunately, it is usually not possible to come up with a combination of orientations that are perfectly evenly distributed in 3D space. This can be understood from the surface of a soccer ball, which consists of a mixture of hexagons and pentagons (which means that the distances between the surface points are not uniform). Therefore, we need to resort to an optimization procedure using, for example, electronic repulsion models. Tables showing gradient combinations for optimized distributions for the given number of orientations can be found elsewhere (see the literature list at the end of this chapter).

6.8.3   The number of gradient orientations

Figure 6.12 shows a comparison between two different types of protocols for gradient orientations. In this example, two different

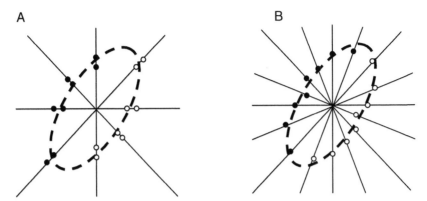

Fig. 6.12 Comparison between repeated measurement with fewer orientations (A) and no repetition with more orientations (B). For simplicity, fitting of a 2D oval is demonstrated, which requires at least three orientations to define the shape and orientation. In (A), a four-orientation measurement is repeated twice, (a total of eight points), which leads to two diffusion constant values along the four measurement axes (black dots). In (B), the eight-point measurements are made along eight evenly distributed axes. Because we assume a symmetric oval for the fitting, the measurements in the reflected side (white dots) are not necessary.

types of eight-point measurements are compared. In Fig. 6.12A, apparent diffusion constants along four different orientations are measured twice. In Fig. 6.12B, the eight measurements are used for eight different orientations. As long as the orientations are evenly distributed, these two types of measurement should provide comparable SNR (note that we need at least six orientations for 3D cases). There are many simulation results published for this type of comparison, and some suggest that the latter approach provides higher SNR or noise is more evenly distributed in the space. However, it is not clear whether the possible improvement, if any, is significant enough to be detected in real studies, in which noise and reproducibility may be dominated by patient motion and the stability of the scanners. Also, repeated datasets (Fig. 6.12A) may have a benefit for quality control of DTI data. For example, if we have more than three repetitions (three redundant datasets), we can calculate standard deviations of the repetitions. If there are coregistration errors or ghosting, they would lead to large standard deviations.

6.8.4 Which protocol should we use?

If we have time to acquire more than six gradient orientations, an important question is whether we should use the time for different b-values, repeated measurement (signal averaging) or a larger number of gradient orientations. This is a complicated issue because there are various factors to consider. First, the most important question is whether we get different results by using different schemes. For example, it is known that signal decay by diffusion weighting is not linear in high b-value ranges (typically more than $b = 3000\,\mathrm{s/mm^2}$, Fig. 6.13). If such high b-values are used, the diffusion constants and possibly diffusion anisotropy would be different from those obtained with lower b-values.

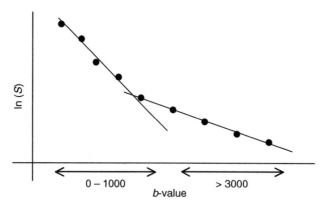

Fig. 6.13 A schematic diagram of nonlinearity of signal decay in a high b-value range. It is known that the slope of signal decay becomes less steep when the b-value is larger than $3000\,\text{s/mm}^2$. In high b-value ranges, measured diffusion constants rely on the choice of b-values.

We do not expect that this type of bias occurs due to choice of the number of gradient orientations or the number of measurement repetitions.

Second, it is possible that we get different SNR, depending on imaging protocols. However, there may not be a single protocol that can simultaneously optimize all imaging parameters for all brain regions. The best b-values or range of b-values may vary, depending on brain regions with different anisotropy or depending on which diffusion parameters we are interested in (diffusion constant, anisotropy, or fiber orientation).

Third, there may be other dominant factors that influence SNR or measurement reproducibility. It is highly likely that some brain regions have a large degree of pulsation motions (e.g., around ventricles), which dominates measurement errors. In such a case, a small degree of SNR improvement by "better" b-value schemes may not be practically important. Another example is echo time: some gradient combinations or smaller b-values allow us to use shorter echo time. Even if such combinations may not be evenly distributed in 3D space, the gain in SNR by using shorter echo times may overcome the benefit of uniform sampling.

6.8.5   Protocol setup flowchart

In conclusion, we do not have a consensus about the "best" diffusion-weighting protocol. However, there are schemes that are currently used. These are: (1) the use of only one b-value at around $600–1200\,\text{s/mm}^2$; (2) the avoidance of using highly skewed gradient orientations; and (3) the use of more orientations is often preferred, but repeated measurements (two to three times) also have a merit. Based on these considerations, an example of decision making for a protocol setup is shown below:

1) Decide the scanning time that can be allocated to DTI, which will determine the number of DWIs;

- With 3 min of scanning time, approximately 30 DWIs can be recorded with single-shot EPI, 96×96–128×128 imaging matrix, and 40–50 slices.
- With typical 1.5T scanners with gradient strength of more than 4 G/cm, 60 DWIs with 2.5 mm isotropic resolution should provide decent SNR.

2) Decide the b-value strength;

- Typically, 600–1200 s/mm^2 is used.

3) Decide the number of gradient orientations;

- If six orientations are used, make sure this is evenly distributed or b-value efficient (shorter TE). For example, the combination $(1, 1, 0)$, $(1, 0, 1)$, $(0, 1, 1)$, $(1, -1, 0)$, $(1, 0, -1)$, $(0, 1, -1)$ is not evenly distributed but is b-value efficient.
- Many current scanners support 12–32 orientations. Any of these combinations should work.

4) Add $b0$ images with 1:6–1:10 ratio for $b0$:DWI;
5) Divide the number of total images by the number of images for one complete dataset (the number of $b0$ + the number of gradient orientations). This gives the number of repetitions. For example, if 60 DWIs are acquired with a 12-orientation scheme, the number of repetitions is 5;
6) Save the repeated measurements into separate files instead of using real-time signal averaging by the scanners;
7) Perform image quality inspection if possible;
8) Perform rigid-body coregistration if brain positions have shifted. Perform eddy current correction using linear affine transformation, if necessary; and
9) Perform tensor calculation.

Chapter 7

New image contrasts from diffusion tensor imaging: theory, meaning, and usefulness of DTI-based image contrast

7.1 TWO SCALAR MAPS (ANISOTROPY AND DIFFUSION CONSTANT MAPS) AND FIBER ORIENTATION MAPS ARE IMPORTANT OUTCOMES OBTAINED FROM DTI

Once we obtain the six parameters of the diffusion ellipsoid at each pixel, our next task is to visualize it so that we can appreciate the neuroanatomy. The most complete way of doing this is to place the 3D ellipsoid at each pixel. However, this is not really a practical method for routine use. First of all, ellipsoids become too small to see at each pixel unless a small region of the brain is magnified. Even if they are magnified, there are such freedoms for viewing as view angle, light source, and shading. Unfortunately, our eyes (or the computer screen) can effectively appreciate (or display) only images with 8-bit (256) grayscale (i.e., pixel intensity) or 24-bit (red/green/blue, RGB) color presentation. Therefore, it becomes important to reduce the six-parameter/pixel information to 8-bit grayscale or 24-bit color presentation. The former method can represent only one 8-bit parameter and the latter up to three 8-bit parameters simultaneously by assigning R, G, and B to each of them. The two most commonly used images with grayscale are the average apparent diffusion constant (average ADC) and the aniso-tropy map. There is some confusion about the definition of average ADC, but it is common to use one-third of trace value ($(\lambda_1 + \lambda_2 + \lambda_3)/3$) for the presentation (also often called simply "trace"). Please refer to Fig. 5.1. Trace values are widely used because it is insensitive to the orientation of the fibers (in this example, the trace is always 3.5). There are many ways to present the extent of anisotropy. The easiest way is to take the ratio of the longest and shortest axes (λ_1/λ_3). The more elongated the ellipsoid, the larger the number becomes. Unfortunately, this simple metric is known to be very sensitive to measurement noise. A more elegant way is, for example, to use metrics related to the difference of the three parameters λ_1, λ_2, and λ_3; $(\lambda_1 - \lambda_2)^2 + (\lambda_1 - \lambda_3)^2 + (\lambda_2 - \lambda_3)^2$. This index is 0 for a sphere ($\lambda_1 = \lambda_2 = \lambda_3$) and becomes larger as the ellipsoid deviates from a sphere. Many indices for anisotropy used nowadays

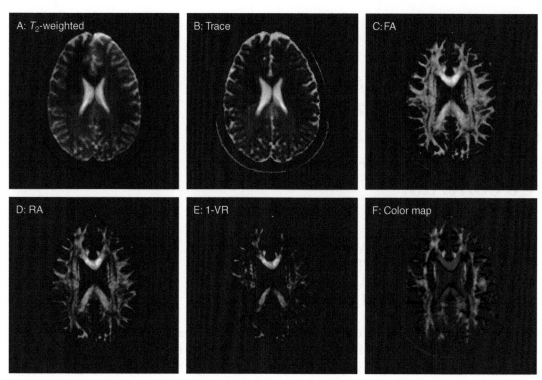

Fig. 7.1 Various image contrasts obtained from DTI. A: reference T_2-weighted image (least-diffusion-weighted image), B: trace map, C: fractional anisotropy map, D: relative anisotropy map, E: volume ratio map, and F: color-coded orientation map. In the color-coded map, red, green, and blue represent fibers running along the right–left, anterior–posterior, and superior–inferior axes, respectively.

are scaled between 0 and 1. One of the most widely used indices, fractional anisotropy (FA), is (Fig. 7.1):

$$\mathrm{FA} = \sqrt{\frac{1}{2}} \frac{\sqrt{\left((\lambda_1 - \lambda_2)^2 + (\lambda_2 - \lambda_3)^2 + (\lambda_3 - \lambda_1)^2\right)}}{\sqrt{\lambda_1^2 + \lambda_2^2 + \lambda_3^2}} \qquad (7.1)$$

The ADC and anisotropy-based maps are scalar maps and can be visualized in the same way as conventional MRI using the 8-bit grayscale. The same quantification approaches such as manual region of interest (ROI)-drawing can be used to quantify them.

7.2 SCALAR MAPS (ANISOTROPY AND DIFFUSION CONSTANT MAPS) AND FIBER ORIENTATION MAPS ARE TWO IMPORTANT IMAGES OBTAINED FROM DTI

So far, we have been focusing on scalar values derived from eigenvalues (λ_1, λ_2, λ_3). Eigenvectors, on the other hand, carry orientation information, which is less straightforward to visualize, quantify, and interpret. In many studies, we discard v_2 and v_3 and concentrate on the orientation of the vector, v_1, which is assumed to represent the local fiber orientation.

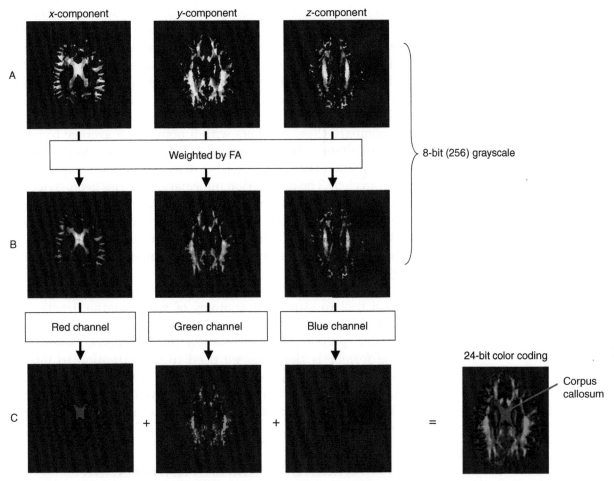

Fig. 7.2 Steps to create color-coded fiber ($\mathbf{v}_1$) orientation maps. The x, y, and z components are obtained from the unit vector $\mathbf{v}_1$ ($\mathbf{v}_1 = [x, y, z]$).

One of the most popular ways to visualize the orientation information is with color-coded maps (Fig. 7.1F). The $\mathbf{v}_1$ is a unit vector. It consists of x, y, and z components ($\mathbf{v}_1 = [x, y, z]$) that fulfils $x^2 + y^2 + z^2 = 1$ and each scaled within the 0–1 range ($0 \leq x \leq 1$, $0 \leq y \leq 1$, $0 \leq z \leq 1$). These x, y, and z components can be presented separately using the grayscale as shown in Fig. 7.2A, in which 256 (8 bit) steps of a grayscale is assigned to each vector component. To mask low-anisotropy regions where there are supposed to be no dominant fibers, these vector component images can be multiplied by an anisotropy map such as FA, which produces cleaner and more informative images. These FA-weighted vector-component images can be presented by grayscale images, as shown in Fig. 7.2B. However, it is not an easy task to appreciate fiber orientation information using three separate images. To better visualize fiber orientations in one image, a 24-bit color presentation, which uses RGB (8-bit each for red, green, and blue) channels, has been postulated. Namely, the x, y, and z component images are assigned to three RGB principal colors and combined to make one color-coded map (Fig. 7.2C). For example, when the $\mathbf{v}_1$ is [1, 0, 0] (right–left orientation), the RGB channels receives [255, 0, 0], which looks pure red. When $\mathbf{v}_1$ is $[1/\sqrt{2}, 1/\sqrt{2}, 0]$, the RGB channel is [181, 181, 0], which is yellow. It requires a bit of practice, but it should not take long to familiarize yourself with this color scheme and start reading the orientation information presented by colors.

The fiber orientation is unique information; no conventional MR images can provide similar information. If an ROI is drawn on a scalar map, we can obtain information such as T_2, ADC, or FA. We can compare these numbers with a population average to judge if they are abnormally high or low. Although we can do the same for a $\mathbf{v}_1$ map and obtain average fiber orientation, its interpretation is far from clear. The orientation information carries geometrical (i.e., anatomical) information, which requires a fundamentally different approach to appreciate and quantify. If there is a group of pixels that share similar fiber orientations, those pixels are part of a specific white matter tract. Then the number of pixels or the way they cluster reflects the size and shape of the tract. Similar to studying the cortical geometry using T_1-weighted images, we now have a tool to study the anatomy of white matter tracts. When we use DTI, we should know whether we are interested in pixel intensity (photometric) or anatomy (morphometric) information. If the former is the case, we need to pay attention to intensity values of scalar maps such as FA and ADC maps. If the latter is the case, we are interested in sizes and shapes of structures inside the white matter, which are revealed by characteristic fiber orientations.

7.3   THERE ARE TUBULAR AND PLANAR TYPES OF ANISOTROPY

So far, we have looked at only FA as a measure of diffusion anisotropy. There are many other ways to represent the anisotropy; widely used anisotropy measures include:

$$\text{RA} = \sqrt{\frac{1}{2}} \frac{\sqrt{(\lambda_1 - \lambda_2)^2 + (\lambda_2 - \lambda_3)^2 + (\lambda_1 - \lambda_2)^2}}{\lambda_1 + \lambda_2 + \lambda_3}$$

$$\text{VR} = \frac{\lambda_1 \lambda_2 \lambda_3}{\left((\lambda_1 + \lambda_2 + \lambda_3)/3\right)^3}$$

where RA stands for relative anisotropy and VR for volume ratio. These indices are all scaled from 0 to 1, although contrast of the VR is inverted. Figure 7.1 shows a comparison of different anisotropy maps.

Up to now, we have assumed that image pixels in the white matter have a single dominant fiber population with a coherent fiber orientation (Figs. 7.3A and 7.3F(1)). If this is the case, the length of the diffusion ellipsoid along the longest axis (λ_1) should be much longer than those of the other two axes (λ_2 and λ_3); i.e., $\lambda_1 > \lambda_2 \approx \lambda_3$. There are situations, however, where $\lambda_1 \approx \lambda_2 > \lambda_3$ (Figs. 7.3B, 7.3C, and 7.3F(2)). Let us call the former ellipsoid the "tubular ellipsoid" and the latter the "planar ellipsoid." When the

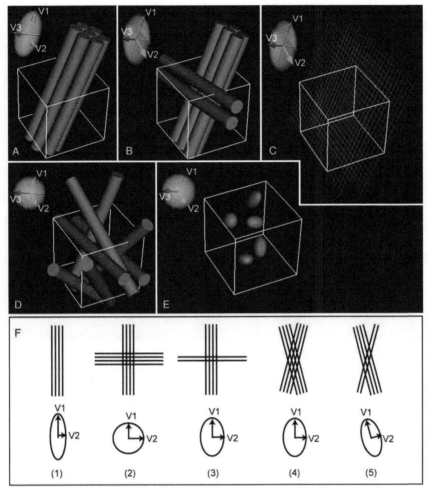

Fig. 7.3 The shape of diffusion tensors and local tissue structures. A: If there is only one group of fibers (blue cylinders for fibers) inside an imaging voxel (the cubic frame), the diffusion tensor has a tubular shape. B–C: If there are two crossing fiber groups (B) or barriers with low permeability (C, blue wireframe surfaces), the diffusion tensor has a planar shape. D–E: For more complex local tissue structures, for example, multiple crossing fiber groups (D) or no fibers (E), the diffusion ellipsoids become spherical. The diffusion tensors were visualized as gray ellipsoids. F: Various configurations of crossing fibers and corresponding 2D diffusion ellipsoids. The number of lines represents the relative population of fibers in each group (reproduced from Zhang et al., *Magn. Reson. Med.* 55, 439, 2006 with permission). This figure was provided courtesy of Dr. Jiangyang Zhang, Johns Hopkins University.

ellipsoid is planar, we cannot draw conclusions about the underlying neuroanatomy because different anatomical configurations may lead to the same planar shape as shown in Figs. 7.3B, 7.3C, and 7.3F(2–5). When the ellipsoid is tubular, the second and third eigenvector orientations are degenerated, and when the ellipsoid becomes completely planar (Fig. 7.3F(2)), the first and second eigenvectors degenerate, meaning the first eigenvector does not represent fiber orientation. Even if $\lambda_1 > \lambda_2 > \lambda_3$ and all eigenvectors are not degenerated, the second and third eigenvectors (v_2 and v_3) do not necessarily represent any fiber orientations, as illustrated in Fig. 7.3F(4 and 5). Note that none of the anisotropy measures described earlier (FA, RA, and VA) differentiate tubular and planar types of anisotropy modes (both have high anisotropy).

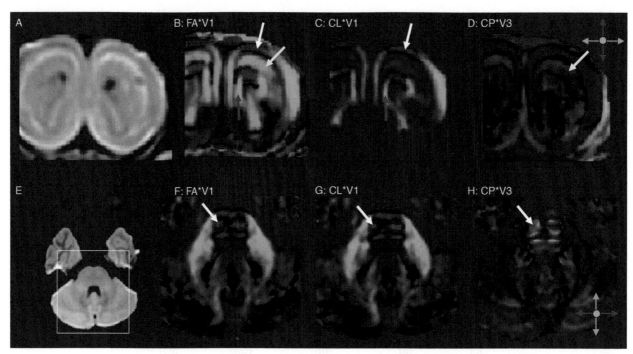

Fig. 7.4 Comparison of FA, CL, and CP maps of mouse embryo (A–D) and the pons of human brains (E–H). Images in (A) and (E) are isotropically diffusion-weighted images for anatomical guidance. FA (B, F) and CL (C, G) maps are color coded by v_1 orientations, while CP maps are color coded by v_3. In the mouse brain, white, yellow, and pink arrows indicate the cortical plate, intermediate zone, and periventricular zone. In the human brain, white arrows indicate the corticospinal tract. Images were reproduced from Zhang et al., *Magn. Reson. Med.* 55, 439, 2006 with permission. This figure was provided courtesy of Dr. Jiangyang Zhang, Johns Hopkins University.

It is possible to devise contrast mechanisms that can differentiate the tubular and planar types of diffusion anisotropy. For example, among the simplest indices are Westin's CL and CP:

$$CL = \frac{\lambda_1 - \lambda_2}{\lambda_1} = 1 - \frac{\lambda_2}{\lambda_1}$$

$$CP = \frac{\lambda_2 - \lambda_3}{\lambda_1}$$

The CL contrast becomes bright for tubular ellipsoid regions but remains dark for planar regions. The CP becomes bright only for planar regions. For both indices, isotropic regions remain dark. Figure. 7.4 compares FA, CL, and CP indices in a mouse embryo and in the human brain. In FA maps, all anisotropic regions, regardless of shape (tubular or planar), are bright. In the CL maps, the planar anisotropic regions are suppressed. We can consider that all bright regions have a relatively simple axonal architecture, with only one dominant fiber population. The difference between FA and CL is surprisingly small in human brains, suggesting that most pixels consist of one dominant fiber. On the other hand, the mouse embryo has two distinctive anisotropy patterns: the cortical plate (white arrow) and periventricular zone (pink arrow) has tubular anisotropy, while the intermediate zone (yellow arrow) has planar anisotropy. This is a good example, which demonstrates that the combination of CL and CP carries more anatomical information than FA alone. The reason for the planar anisotropy of the intermediate zone is most likely due to a mixture of radial glia (columnar orientation that dominates the cortical plate and periventricular zone) and growing axon (tangential orientation), which leads to a tensor depicted in Fig. 7.3F(2) or F(3). Careful inspection of the human brain also reveals that the corticospinal tract (white arrow) has planar properties. Because of the information degeneration issues explained in Fig. 7.3, the CL and CP contrast cannot determine the exact architecture of axons, but they do carry extra information that we cannot always appreciate from FA, and it is worth investigating its values.

7.4   DTI HAS SEVERAL DISADVANTAGES

DTI has many disadvantages compared to relaxation-based MRI. First of all, it is extremely motion sensitive. Diffusion images are sensitive to water diffusion that is in the order of 5–10 μm within the measurement time. Even brain pulsation motion can be larger than this. Unless we use single-shot (SS)EPI, we often get images full of ghosting. While many more elaborate techniques for data acquisition and post-processing have been proposed, the SS-EPI is, and will be, the method of choice in many clinical studies in the foreseeable future due to its robustness. This means, DTI has all the disadvantages of the SS-EPI, which include image distortion and low spatial resolution. Even with the SS-EPI, motion-related ghosting is not uncommon. Second, DTI requires at least seven (in practice, many more) images to perform the tensor fitting. It has all the problems associated with the fitting process, most notably, long imaging time, low SNR, and misregistration due to patient motion. Third, DTI has problems caused by the applications of strong diffusion-weighting gradients. These include long TE (thus, low SNR) and image distortions due to instrumental imperfection (eddy current). Given these disadvantages, it is important to test if DTI provides unique and useful information that cannot be obtained by conventional MRI.

To address this issue, we sometimes quantify anisotropy in brain regions where T_2-weighted images appear normal (normal-appearing white matter) and conclude that "anisotropy could detect significant differences in T_2-normal-appearing white matter". However, a comparison between visual and statistical assessment could be misleading. We have to be very careful because this finding does not necessarily mean that anisotropy provides information that is unique and useful. Depending on SNR and the type of the abnormality (diffuse or concentrated), our eyes may not be able to detect statistically significant changes (by group average) of even 10% in individual cases. If we draw ROIs in such areas and perform statistical analysis, we may detect a significant alteration of anisotropy as well as T_2. It is always recommended to perform careful quantification of relaxation-based parameters to judge the uniqueness and usefulness of DTI parameters.

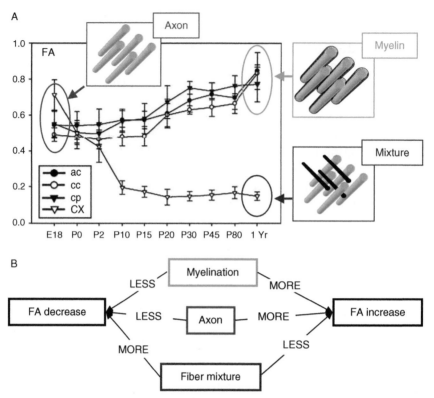

Fig. 7.5 Change of anisotropy in the cortex and white matter tracts of developing mouse brains (A). The mouse brains were perfusion fixed at various developmental stages and scanned by a 11.7T MRI magnet. The letter E and P in the *x*–axis stand for embryonic (E) and postnatal (P) days. The schematic diagram in (B) shows various factors that affect FA values. Abbreviations are; ac: anterior commissure, cc: corpus callosum, cp: cerebrum peduncle, and CX: cortex. Figure (A) is reproduced from Mori and Zhang, *Neuron*, 51, 527, 2006, with permission.

7.5   THERE ARE MULTIPLE SOURCES THAT DECREASE ANISOTROPY

As mentioned in the previous section, there are two important factors we should consider when we perform DTI: first, whether a DTI-derived parameter can provide unique information that cannot be obtained from conventional MR; and second, whether the unique information is useful. This is indeed the case for ADC, which can uniquely detect an early phase of ischemic events, which is surely useful for patient management. Anisotropy is not that simple. It can differentiate gray and white matter, but so can T_1 and T_2. Anisotropy decreases by edema, but T_2 can detect edema too. It has been believed that anisotropy changes by loss of myelin, but again so does T_2.

Previously, myelin was attributed to the source of anisotropy, but there is mounting evidence that nervous tissue can have high anisotropy without myelin. Axon itself is sufficient to confer water diffusion anisotropy, although myelin can still be one of the sources. It is likely that the higher the axonal density is, or the more myelinated the axons are, the higher the anisotropy becomes. In addition to these microscopic contributions, we also know that homogeneity

of axonal orientation within a pixel has a major impact on anisotropy (see also Chapter 4.6 for more discussion). The more the
orientation mixture we have, the lower the anisotropy becomes.
Figure 7.5 shows a good example of sensitivity of FA to multiple
biological events, in which FA of the cortex and white matter
tracts was measured during brain development of C57/BL mice.
Before birth (e.g., E18 in Fig. 7.5), when most of the brain is not
myelinated, both the cortex and white matter tracts already
have high anisotropy. After birth (P0) myelination starts both
in the cortex and white matter tracts, but their time courses
are polarized. FA of the white matter tracts increases
probably due to myelination, and FA of the cortex decreases
probably due to loss of coherent structure. If we can assume that
postnatal increase of FA in white matter is solely due to myelination, we can conclude that approximately 60% of anisotropy is
caused by axonal bundles and 40% is by myelination in adult
brains. However, the assumption may not be true because axonal
density and caliber may also increase in the postnatal stages. In
recent studies by Song et al., a mouse pathological model was used
to compare FA of normal and dysmyelinated tracts and 20%
reduction of FA was found (for a more thorough analysis of cellular
mechanisms of anisotropy, see the review paper by Beaulieu (*NMR
Biomed.*, 15, 435–455, 2002)).

The cortex initially has high anisotropy due to neatly aligned
columnar structures, which later disappears, most likely due to
growth of weblike dendrite networks. The cortex has axons and
more myelin in later stages, but anisotropy decreases. Namely, in a
microscopic (cellular-level) point of view, the cortex should have
higher anisotropy, which is counteracted by macroscopic (pixel-
level) disorganization within a pixel. This macroscopic factor (alignment of fibers within a pixel) is the more dominant factor in DTI-
based anisotropy determination and can override microscopic
factors. This also means that diffusion anisotropy has dependency
on image resolution: the higher the resolution is, the less orientation heterogeneity there is. The sizes of prominent white matter
tracts are often larger than pixel sizes, and dependency on image
resolution may be limited to the boundary of two fibers where partial volume effect occurs. The anatomic scale of fiber mixtures in
the gray matter is much smaller than commonly used imaging resolution (1–3 mm). Therefore, differences in imaging resolutions at
realistic scales influence only small tracts and tract boundaries,
which have the similar anatomical scale as the pixel size (1–3
mm). The cerebellum and some fiber-rich gray matter structures
such as the thalamus may have fiber structures in this scale range
and, thus, their anisotropy may have significant dependency on
image resolution.

The influence of these two-layered factors (microscopic and
macroscopic) is a unique feature of the anisotropy contrast and
cannot be overstressed. When we observe a decrease in diffusion
anisotropy, we tend to relate the change to microscopic factors
such as myelination and axonal damage, but we should bear in

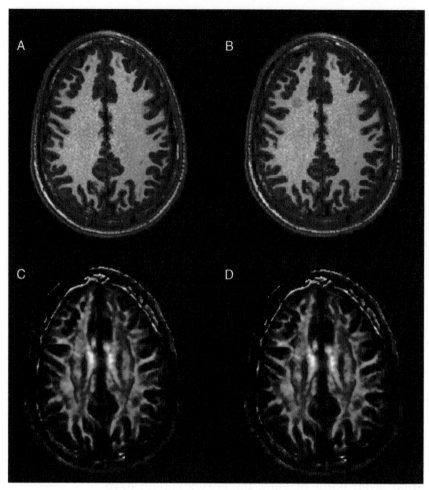

Fig. 7.6 Can you find an abnormal region? In images (B) and (D), a small lesion is artificially made at the same location by reducing the intensity by 40%. Images (A) and (B) are T_1-weighted images, and (C) and (D) are FA maps.

mind that low anisotropy could occur in perfectly healthy white matter if there is macroscopic orientation inhomogeneity. This point can easily be seen in Figs. 7.6A and 7.6C. The white matter looks homogeneous in the T_1-weighted image, while the same region looks very inhomogeneous in the FA map. By shifting an ROI by a small amount, we may find a substantial difference in anisotropy. If we see a difference in FA between control and patient groups, it could be due to a subtle difference in measured areas. Even if we are sure that we are measuring the same anatomical regions, we cannot immediately determine that the measured differences are due to microscopic factors or macroscopic factors (changes in fiber populations).

In a pathological condition, if 20% loss of FA is found, for example, what can we deduce? Compared to relaxation-based contrasts, which are determined by chemical and physiological properties of water molecules, we can conclude that diffusion anisotropy is a dirtier parameter. The sensitivity of anisotropy to multiple sources of physiological and anatomical parameters, and the resultant feature-rich contrast, could be a double-edged sword.

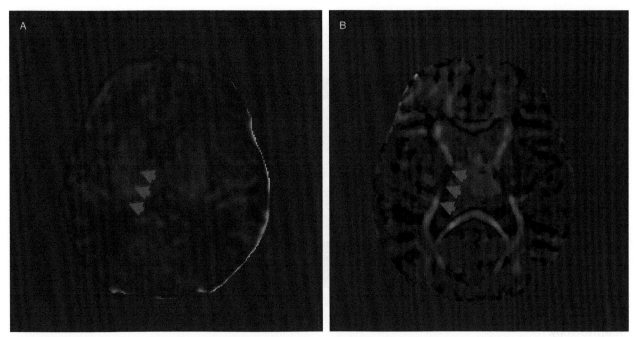

Fig. 7.7 Comparison of a T_2 map (a) and a color-coded map (b) of a 0-month-old subject. The location of the posterior limb of the internal capsule is indicated by pink arrows. It is difficult to identify this white matter tract in the T_2 map due to the lack of myelination.

In Fig. 7.6B and Fig 7.6C, the intensities of a white matter region were artificially reduced by 40%. Such reduction can be readily detected in the homogeneous-looking T_1-weighted image but can be difficult to find in the FA map. We need to resort to right–left asymmetry or nonanatomical patterns of intensity modulation to detect abnormalities.

7.6   ANISOTROPY MAY PROVIDE UNIQUE INFORMATION

Because anisotropy values range from almost 0 to close to 1 in normal white matter, the evaluation of anisotropy values is not straightforward. Dislocation of the ROI by a few millimeters may result in a significant change in anisotropy values. If the lesions can be detected by conventional MR parameters, we may not need DTI. It is often naively assumed that anisotropy is sensitive to "white matter integrity." However, whatever the definition of "integrity" is, relaxation parameters can also be sensitive to loss of the integrity, such as demyelination. One important fact that is obvious from Fig. 7.5 is that anisotropy is sensitive to the existence of the axon itself, while relaxation parameters such as T_2 are not always so. This point can be clearly seen in human neonatal studies (Fig. 7.7).

For example, location of the posterior limb of the internal capsule (indicated by pink arrowheads) can be clearly seen in the color-coded map (Fig. 7.7B) but not in the T_2 map (Fig. 7.7A). This leads to the hypothesis that anisotropy is a unique and sensitive

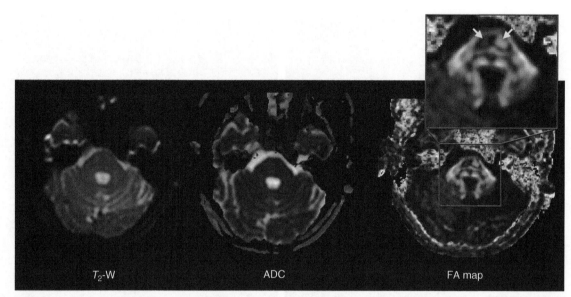

Fig. 7.8 Comparison between T_2-weighted, trace, and FA maps of a patient with stroke-related lesions in the corticospinal tracts at the pons (indicated by yellow arrows). The right corticospinal tract has apparent loss of anisotropy.

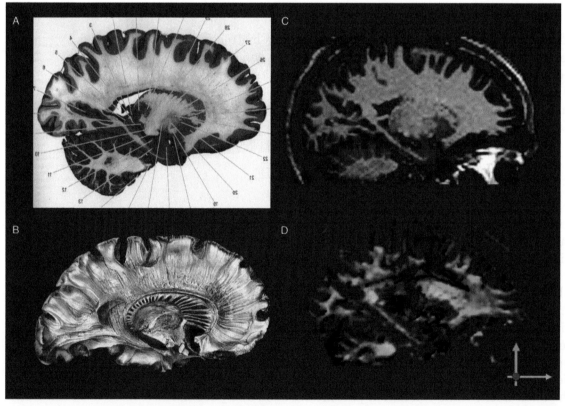

Fig. 7.9 Comparison between histology preparations (A and B), T_1-weighted image (C), and DTI-based orientation map (D). The picture (A) is a simple brain slice of a postmortem sample, while the sample (B) was specially prepared to reveal complex white matter structures. In the color map (D), red, green, and blue color represent fibers running along the right–left, anterior–posterior, and superior–inferior axes. (Copyright-protected materials (A: The Brain Atlas: A Visual Guide to the Human Central Nervous System, Fitzgerald, and B: Atlas Cerebri Humani, S. Kager AG, Basel) are used with permission.)

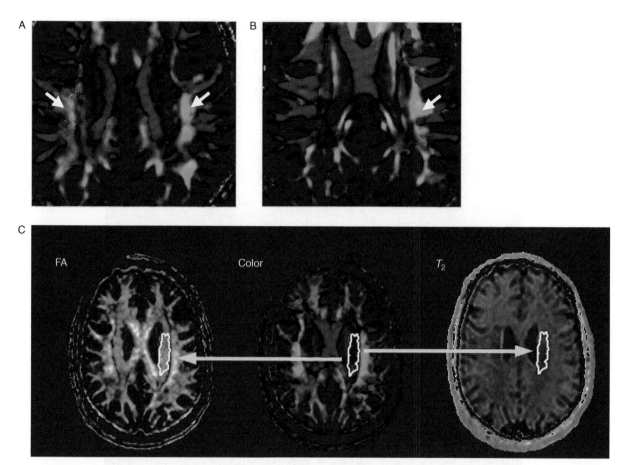

Fig. 7.10 Examples of how to use orientation information in application studies. For qualitative usage, we can use it to visually detect abnormalities in white matter anatomy revealed by the orientation information (A and B). The color maps shown in (A) and (B) are from a healthy volunteer (A) and a patient with displasia (B). It is clear that the green fiber (the superior longitudinal fasciculus running along the anterior–posterior axis) of the left hemisphere of this patient (B) cannot be identified. Another example is to use the anatomical information as a template for quantitative analysis (C). In this example, the coordinates of the corona radiata are manually defined, and the coordinates are superimposed on FA and T_2 maps to quantify them in a tract-specific manner. By counting the number of pixels within the ROI, the cross-sectional area of the fiber can be measured.

indicator of axonal loss, while its specificity to axonal loss is not obvious because it may also be sensitive to myelination status and other abnormalities. To test this hypothesis, we need animal models in which axons are selectively damaged while myelination is preserved and *vice versa*. Figure 7.8 shows an example of a human study in which axonal loss is uniquely detected by anisotropy. This patient has a large stroke infarction in the right hemisphere that involves frontal and parietal lobes. This leads to Wallerian degeneration of the right corticospinal tract at the pons level, which can be clearly identified by a decrease of anisotropy. Interestingly, this lesion did not accompany an apparent change in T_1, T_2, and ADC. The space occupied by the right and left corticospinal tract does not seem to be significantly different. Without histology confirmation, accurate pathology cannot be concluded, but this exemplifies the idea that diffusion anisotropy could provide unique information about axonal abnormalities.

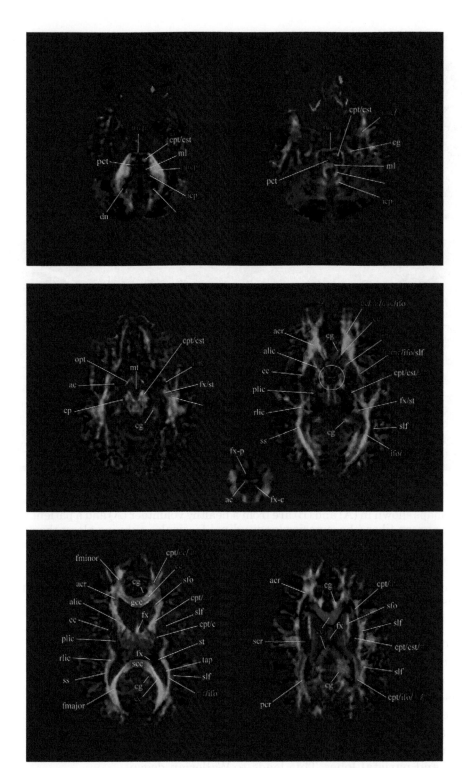

Fig. 7.11 Atlases of human white anatomy delineated by color-coded orientation maps. Eight evenly spaced axial slices are extracted at an interval of 11 mm. Red, green, and blue colors represent fibers running along the right–left, anterior–posterior, and superior–inferior axes. Abbreviations are ac: anterior commissure, acr: anterior corona radiata, alic: anterior limb of the internal capsule, atr: anterior thalamic radiation, cc: corpus callosum, cg: cingulum, cp: cerebral peduncle, cpt: corticopontine tract, cst: corticospinal tract, dn: dentate nucleus, ec: external capsule, fx: fornix, icp: inferior superior peduncle, ifo: inferior fronto-occipital fasciculus, ilf: inferior longitudinal fasciculus, mcp: middle cerebellar peduncle, ml: medial lemniscus, mt: mamillothalamic tract, opt: optic tract, pcr: posterior corona radiata, pct: pontine crossing tract, plic: posterior limb of the internal capsule, ptr: posterior thalamic radiation, rlic: retrolenticular part of the internal capsule, scp: superior cerebellar peduncle, scr: superior corona radiata, sfo: superior fronto-occipital fasciculus, slf: superior longitudinal fasciculus, ss: sagittal stratum, st: stria terminalis, tap: tapatum, and unc: uncinate fasciculus. (Images are reproduced from "MRI Atlas of Human White Matter," Elsevier.)

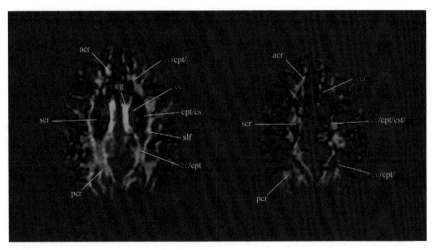

Fig. 7.11 Continued.

7.7 COLOR-CODED MAPS ARE A POWERFUL VISUALIZATION METHOD TO REVEAL WHITE MATTER ANATOMY

The uniqueness and usefulness of the information from anisotropy measurements are not always apparent, and we sometimes do not know the answer until we finish the study. On the other hand, orientation information from DTI is definitely unique. There is no other imaging modality that can provide such information. As mentioned earlier, the fiber orientation provides anatomical, not physiological, information of the white matter. The rich anatomical information carried by a fiber orientation map is clearly shown in Fig. 7.9. Axonal structures in the white matter are not readily appreciable even in postmortem brain slices (Fig. 7.9A). The beautiful sample shown in Fig. 7.9B was obtained by repeated freezing procedures followed by manual dissection. The T_1-weighted image (Fig. 7.9C) looks much like Fig. 7.9A, in which the white matter appears to be a homogeneous field. Apparently, there is a tremendous amount of anatomical information about the white matter in the orientation map (Fig. 7.9D; see also Fig. 7.11).

In Chapter 11, application studies will be divided into two categories, and strategies to use the orientation information will be discussed. The first category is qualitative studies (e.g., Figs. 7.10A and 7.10B). Most radiological diagnosis is based on visual inspection of images. In this type of study, it is important that a new imaging modality be able to provide visually appreciable information about abnormalities. The value of DTI would be even greater if such abnormalities cannot be seen in conventional MRI. This first category includes diseases with significant changes in white matter anatomy, such as developmental problems, Wallerian degeneration, and brain tumor.

The second category is quantitative studies (Fig. 7.10C). In this category, there are two ways in which orientation information could be useful. First, it can provide anatomical templates to refine ROI-based quantitative analysis (photometric studies).

Using color maps, we can identify locations of specific white matter tracts of interest at a given slice level or three-dimensionally. We can, for example, manually define one of the tracts and quantify various MR parameters (tract-specific quantification). We can also semiautomate this process by using a 3D tracking algorithm. Second, it provides information about the size and shape of a specific tract (morphometric studies). While possibilities for these quantification studies are exciting, the actual quantification process is far from straightforward, and we sometimes lack the appropriate tools for it. This point will be discussed in more detail in Chapter 10.

Chapter 8

Limitations and improvement of diffusion tensor imaging

8.1   TENSOR MODEL OVERSIMPLIFIES THE UNDERLYING ANATOMY

MRI is mostly used to detect signals from water molecules. Based on chemical and physical properties of water molecules, various contrasts can be generated such as T_1, T_2, and diffusion-weighted images. We then try to correlate these contrasts to the underlying anatomy. However, the underlying anatomy is always much more complicated than the amount of information we can obtain through MR parameters (or water properties) (Fig. 8.1). We are dealing with a decidedly under-sampled system. To solve this issue, we often use models, simplifications, and assumptions. Let us take T_2 maps. We sample various intensities at different echo times. We usually assume that the T_2 decay is mono-exponential and extract one parameter, T_2. Of course, we know that the actual decay is not mono-exponential because of multiple compartments with different T_2 values (and possible exchange among them). Still, such a simple mono-exponential model serves many clinical and research purposes to locate, say, inflammatory areas. More complicated models often require longer data acquisition times. The amount of clinical and biological information may not increase as much as the scanning time does. The accuracy of the models we use is important, but it should be balanced against practicality and cost–benefit functions.

As mentioned in previous chapters, the tensor model is a bold assumption. The model assumes that brain anatomy consists of a fiber population with a uniform fiber angle within a pixel, which is often not the case. In Fig. 7.3 of Chapter 7, simple schematic 2D diagrams show the relationship between fiber architectures and the results of tensor fitting. No matter what the fiber architecture is, tensor fitting forces the results to be fitted into one diffusion ellipsoid, in which there is only one long axis that is supposed to represent the fiber orientation. This is only true when there is only one fiber population in a pixel (Fig. 7.3F(1)). If there are two fiber populations (Fig. 7.3F(2–5)), different fiber architecture may lead to the same fitting results, and the longest axis may not represent the orientation of any fibers. Two different types of neuroanatomy may lead to the same DTI result (information degeneration). In this case, our measurements are under-sampled, and estimation

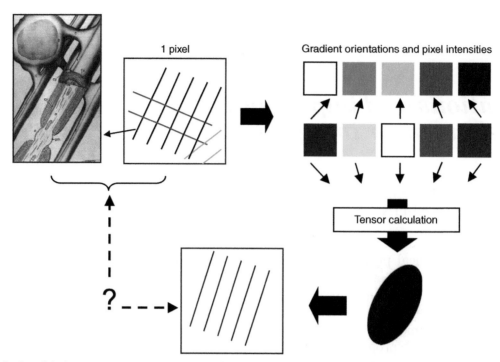

Fig. 8.1 Relationship between neuroanatomy and anatomical information obtained from DTI. Our pixel size for DTI measurements is typically 2–3 mm, in which there could be multiple populations of axonal tracts with different orientations. They could be mixed in a microscopic scale (e.g., black and pink fibers shown in the pixel) or partial volume effect (blue fibers). Each axonal bundle contains many axons and water molecules distributed inside axons, myelin sheaths, astrocytes, or extracellular space. The underlying axonal neuroanatomy is decidedly complicated. Our raw data are diffusion measurement results along many orientations, to which we fit a tensor. From the fitting result, we deduce the underlying axonal structure. Not surprisingly, the deduced anatomy is a first-order approximation of the complicated anatomy. (Artist's drawing of axonal fibers is reproduced from Carpenter's Human Neuroanatomy, Williams & Wilkins, with permission.)

of fiber architecture from diffusion measurement is ill-posed. Depending on the angle and population ratios of fiber populations, diffusion may even look isotropic. The real issue is that we expect many regions in our brain to have fiber architectures like those in Fig. 7.3F(2–5). In such cases, the tensor model simply fails. Is the tensor model wrong then? Again, the answer depends on our biological and clinical questions, in which we have to consider practicality and cost-benefit functions. It is safe to say that the tensor model is the best model for first-order approximation of the water diffusion process and for estimation of neuroanatomy. Depending on research topics, the tensor model may extract nearly 100% of the anatomical information we require and what is left is fitting residuals, which may not be worth spending extra scanning time to retrieve. Conversely, the tensor model could be completely a wrong model. In most cases, our research falls between these two extremes, but the tensor model remains one of the best models with a high cost-benefit function. Ultimately, it is important to know what our biological question is and what the limitations are of the tensor model to answer the question.

Another practical issue is that the amount of information that DTI provides (six parameters/pixel) is already beyond our ability to appreciate. For example, for visualization, we usually reduce the information to one parameter/pixel (e.g., ADC and FA maps) or color-coded maps in which information on two eigenvectors (v_2 and v_3) are discarded. When it comes to quantification, we know how difficult it is to analyze brain anatomy using T_1-weighted images. With six times more information per pixel, the tensor model could already contain too much data for us to analyze. It would be easy to increase the amount of information by acquiring more data or by fitting to more elaborated models. However, we need to consider where the bottleneck is for information extraction in our study and for efficient research design.

8.2 THERE ARE MORE SOPHISTICATED "NON-TENSOR"-BASED DATA PROCESSING METHODS, WHICH REQUIRE DIFFERENT DATA ACQUISITION PROTOCOLS

Let us go back to Fig. 8.1 and think where we lost detailed anatomical information. The first information loss occurs because we are trying to deduce axonal structure using water diffusion. Water diffusion is influenced by many anatomic structures during the measurement time, and it cannot represent the property of any single cell types or subcellular entity. In addition, we need to integrate the water diffusion property over the 2–3 mm of the pixel size. Let us call this "Level 1 information loss." The second information loss occurs because we measure diffusion along only a limited number of orientations. We know that we need only six orientations to calculate the tensor, so we are not very motivated to spend a long time measuring 100s of the orientations. This is "Level 2 loss." The third loss occurs when we perform the tensor fitting. Even if we measure diffusion constants along 60 orientations, we reduce the data into six parameters. This is "Level 3 loss." Level 1 is inherent in diffusion measurement, and we cannot do much about it other than improving the image resolution. What we can do is to try to minimize Level 2 and Level 3 losses.

As mentioned earlier, tensor calculation is a first-order approximation of the water diffusion process. It leaves residuals after the fitting, which are thrown away. It is an important part of the research effort to find out if there is important anatomical information left behind or there is significant inaccuracy in this first-order approach. To solve the oversimplification, we inevitably need more extensive measurements to depict the tissue anatomy (less Level 2 loss). Such alternative approaches require sampling of the diffusion constant along many more axes (i.e., with higher angular resolution) and with higher b-values (see Section 8.3 to find why we need high b-values); when a single b-value is used, the technique is known as high-angular resolution diffusion imaging (HARDI). In the past few years, there has been considerable research on advanced data processing strategies, such as diffusion

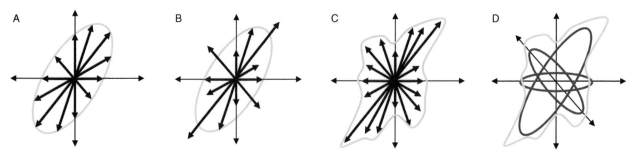

Fig. 8.2 Schematic diagram of non-tensor diffusion properties and models to delineate such systems. For simplicity, a 2D diagram is used. In the tensor model, measurement results are fitted to a diffusion ellipsoid (A). In a real situation, the fitting may not be perfect, not only because of noise and measurement errors but also multiple fiber populations in the pixel (B). In this case, diffusion along many more axes can be measured to accurately delineate the diffusion property (C). In (D), three tensors are fitted to model the complicated diffusion properties.[1]

spectrum imaging (DSI), Q-ball imaging, spherical deconvolution, and persistent angular structure MRI (PAS-MRI). For details of these techniques, please refer to recent papers and review articles listed at the end of this chapter. Here, we will briefly discuss the principle of some of these more elaborate approaches.

In Fig. 8.2A, the idea of DTI is shown, in which a smooth and symmetric 3D ellipsoid (or 2D oval in this 2D analogy) is fitted to multiple measurements of diffusion constants along different orientations. If it fit well, we can describe the diffusion anisotropy by six parameters of the ellipsoid. However, Fig. 8.2B could be the reality. Neuroanatomy and, thus, the water diffusion process are too complicated to be delineated by a simple ellipsoid. To delineate such a complicated diffusion process, we can measure the diffusion constant along many more orientations (Fig. 8.2C). Diffusion spectrum imaging uses this type of multiple measurements (typically 300–500 measurements, acquired at many directions and with multiple b-values up to $\sim 17{,}000$ s/mm^2) to delineate the way water diffuses. Q-ball imaging has been proposed as a simplified version of this approach; in this technique, only the angular information of the diffusion properties is characterized, and this allows the calculation to be performed with less orientation measurements than DSI. Figure 8.3 shows an example of pixel-by-pixel visualization of a Q-ball result. By assuming that each spike of the shapes corresponds to the orientation of an individual fiber population, we can see that some pixels contain multiple fiber populations with different orientations and population ratios.

Spherical deconvolution is an alternative approach that determines the actual distribution of fiber orientations within a voxel directly from the HARDI data, regardless of how many fiber bundles may be present. It is based on the fact that the signal measured as a function of orientation during a HARDI acquisition is the sum of the contributions from the various fiber bundles present. In this way, the volume fractions and orientations of the

[1] Strictly speaking, as mentioned in Chapter 4, the profile of diffusion constants in the space looks like a peanut but not an ellipsoid.

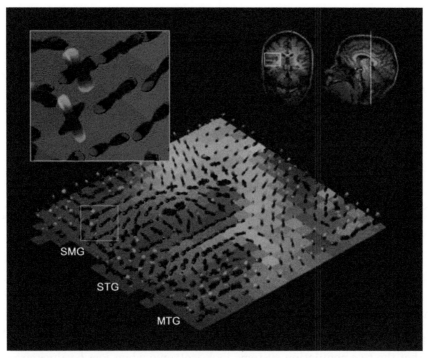

Fig. 8.3 An example of Q-ball imaging. A profile of water diffusion was reconstructed at each pixel from 492-orientation measurement results. Multiple fiber populations can clearly be seen in many pixels. SMG, STG, and MTG stand for supramarginal, superior temporal, and middle temporal gyrus, respectively (the figure is reproduced from Tuch et al. *Neuron*, 40, 885–895, 2003, with permission).

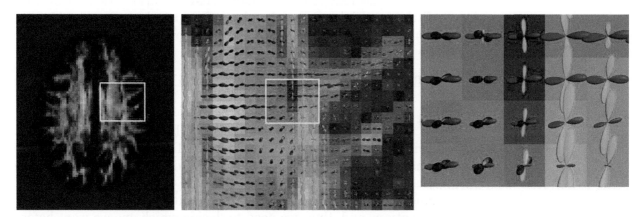

Fig. 8.4 An example of spherical deconvolution based on a 60-orientation measurement. The only major assumption of this technique is the uniform anisotropy of the constituent fibers. Crossing fibers can be resolved with a much smaller number of orientation measurements than DSI. In this example, areas with two and three fiber populations can be resolved (images courtesy of Drs. J.-D. Tournier and F. Calamante).

various fiber bundles present can be estimated. An example is shown in Fig. 8.4.

Multiple-tensor fitting is another option to quantitatively analyze the multiple fibre populations (Fig. 8.2D). In this example, three tensors are used for the fitting. Because each tensor requires six parameters (18 parameters in total), and we also need population ratios of the three tensors, this model would be based on a 20-parameter fitting. Theoretically, we can solve this model if we

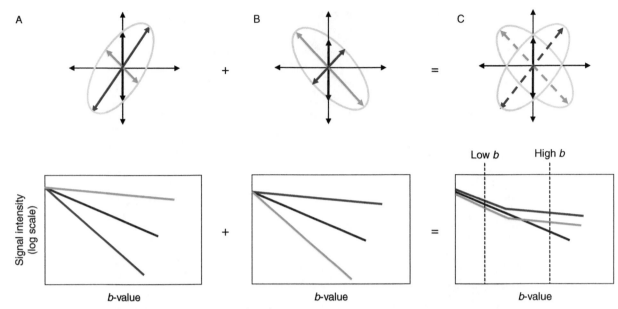

Fig. 8.5 A schematic diagram of fiber population and hypothetical signal attenuation. (A) and (B) show situations where there is only one fiber population. Colors of gradient axes correspond to colors of linear signal decay. If these two fiber populations coexist, we expect superimposition of these two signal decay results (C). Note that the two-population systems look isotropic when low b-values are used. Information about anisotropy with two fiber populations is revealed only in the high-b-value regions. In the graphs, signal intensities are in log scale.

have measurements along more than 20 orientations. However, fitting multiple tensors has been found to be very unstable and unreliable for N (the number of fiber populations) > 2.

8.3   NON-TENSOR MODELS USUALLY REQUIRE HIGH b-VALUES

If we have a dataset consisting of a large number of gradient orientations, we can perform both DTI and non-DTI image analyses. Is it then a good idea to acquire DTI data with a large number of gradient orientations all the time? Unfortunately, DTI and non-DTI data acquisition are not quite compatible in practical situations and, thus, we need to decide which one we want before acquiring the data. This is because non-tensor analyses require not only a large number of measurement orientations but also large b-values. The b-values typically used for DTI (about 1000 s/mm^2) are not optimal for non-tensor models. The schematic diagram in Fig. 8.5 explains the reason for this. When there is only one fiber population (A or B), we expect mono-exponential decay of signal intensity as a function of b-value. For Fig. 8.5A, we expect larger signal attenuation when the gradient is applied along the red-colored axis and smaller attenuation along the green-colored axis. This result should be inverted for (B). When there are two fiber populations (and if there is no water exchange between the two compartments within the imaging time), we expect superimposition of the two results of (A) and (B). When gradients are applied along the red or green axes, the resultant signal decay becomes biexponential; namely the addition of fast- and slow-decaying components. Along the blue-colored axis, the decay remains mono-exponential. Interestingly, when the b-value is small, these three decaying lines cannot be distinguished, and the system appears to be an isotropic medium. Anisotropy information manifests itself only with higher b-values. The higher the b-value, the more discrimination power we can obtain. Of course, if the b-value is too high, signals reach the noise floor, and we lose the power again. For *in vivo* human brains, $b > 5000$ s/mm^2 is often used to achieve this "high" b-value range.

In a sense, the non-tensor model is superior to the tensor model because it carries more anatomical information and, if required, it can also produce tensor-based results. However, there are disadvantages in these approaches in practical situations. First, it requires a large number of gradient orientations. With a 96 × 96 in-plane matrix and 50 slices, acquisition of 30 DWIs should take approximately 3 min by SS-EPI. Three hundred orientations, thus, would require an imaging time of 30 min. The second issue is the adoption of high b-values. The higher b-value requires longer echo time. To obtain sufficient SNR with high b-values and long echo time, we may need to reduce the resolution or resort to more signal averaging (thus, longer scanning time). Within the same amount of time, we could enhance resolution if we use simpler diffusion tensor imaging, thus reducing the Level 1 loss; namely, while non-tensor approaches increase the amount of information within a pixel (decreased Level 2/3 loss), DTI could increase the amount of anatomical information of the entire brain by reducing the pixel size within the given amount of time (less Level 1 loss). High b-values also lead to more motion artifacts and eddy current distortion. The resultant noisy DWIs make it harder to perform motion correction and to judge the amount of artifacts.

In conclusion, both tensor and non-tensor methods have advantages and disadvantages. The method of choice depends entirely on what our biological questions are. If the question requires us to resolve crossing fibers within a pixel and the crossing occurs at the microscopic level, the non-tensor approach would be the method of choice. If DTI is used without sufficient knowledge about its limitations, the research could lead to wrong biological conclusions. It is, however, important to realize that, no matter which method we use, the underlying anatomy could still be far more complicated than what we can deduce from water motions, i.e., the Level 1 loss always prevails.

Chapter 9

Three-dimensional tract reconstruction

9.1 THREE-DIMENSIONAL TRAJECTORIES CAN BE RECONSTRUCTED FROM DTI DATA

In Chapter 7, I emphasized the usefulness of the color map in revealing intra-white-matter architectures (white matter tracts) based on pixel-by-pixel fiber orientation information. However, the 2D-based color map can only reveal a cross section of white matter tracts, which often has convoluted structures in the 3D space, and it is difficult to appreciate their 3D trajectories from the slice-by-slice inspection. Computer-aided 3D tract tracking techniques (called tractography) can be very useful to understand tract trajectories and their relationships with other white matter tracts and/or gray matter structures. For example, in Fig. 9.1A, a color map of the pons is shown. It is easy to see the trajectory of the middle cerebellar peduncle (mcp) from its color. However, 3D reconstruction (Fig. 9.1B) shows that the superior part of the mcp projects to inferior–lateral areas of the cerebellum, while the inferior part projects to superior–medial areas of the cerebellum, which agrees with the anatomist's drawing of white matter anatomy of the brainstem (Fig. 9.1C). This type of anatomic information can be much more easily appreciated from the 3D reconstruction.

9.2 THERE ARE TWO TYPES OF RECONSTRUCTION TECHNIQUES

Roughly speaking, there are two types of tract reconstruction techniques. One is based on "tract propagation," and the other is based on "energy minimization." The former strategy is a deterministic approach because it provides only one solution (trajectory) from a given seed, and there is no *a priori* knowledge about the destination of the propagation. In the latter approach, two arbitrary points in the space can be selected, and the methods provide the most probable path to connect together the two points with likelihood of connection. A detailed description of these techniques is beyond the scope of this book; here, the concepts of simple tract propagation methods are described.

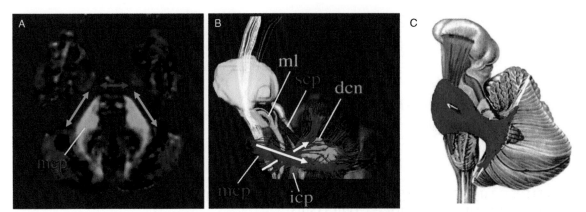

Fig. 9.1 Comparison with a 2D color-coded orientation map (A), 3D tract reconstruction result (B), and hand-drawn figure of the brainstem by an anatomist (C). In the color map (A), red, green, and blue represent the fiber orientation at each pixel (red: right–left, green: anterior–posterior, blue: superior–inferior). Approximate fiber orientation of the middle cerebellar peduncle (mcp) is shown by arrows. The 3D tract reconstruction result reveals 3D trajectory of the mcp, which is colored red. As indicated by two white arrows, the mcp, which has a sheet-like structure, is twisted more than 90° as it travels toward the cerebellum. This result agrees well with the anatomist's drawing (C) and is very difficult to appreciate from slice-by-slice inspection of 2D color maps. Abbreviations are dcn: deep cerebellar nucleus, icp: inferior cerebellar peduncle, ml: medial lemniscus, and scp: superior cerebellar peduncle. (Anatomist's drawing reproduced from "The Human Central Nervous System," Springer-Verlag, with permission.)

9.3   THERE ARE THREE STEPS IN THE TRACT PROPAGATION MODELS

In the tract propagation approach, there are three important steps. First, we have to estimate local fiber orientation, in which there are two options; one is the tensor (Fig. 9.2A, diffusion ellipsoid), and the other is the vector of the longest axis of the ellipsoid (Fig. 9.2B). The latter approach is more straightforward, but we inevitably discard some information when the tensor is converted to a vector. For example, reliability of the vector orientation in low-anisotropy regions will be lower than in high-anisotropy regions. In an extreme case, the orientation of the longest axis is not reliable at all if the ellipsoid is isotropic or has a disk shape, in which the longest axis is determined by noise (the pixel with an asterisk in Fig. 9.2).

The second step is propagation of a line based on the tensor or vector information. There are many postulated methods depending on interpolation and simulation approaches, as discussed later in this chapter. The third step is termination of the propagation (Fig. 9.3). The two most often used criteria are low-anisotropy (Fig. 9.3A) and sharp bend (Fig. 9.3B). In low-anisotropy regions, we can assume that there is no coherent fiber population, and, for the sharp bend, image resolution is not high enough to reliably follow the tract. The latter threshold also prevents the tracking from jumping to adjacent unrelated tracts. Because the gray matter typically has an FA of 0.05–0.15, it is common to use FA > 0.15–0.3 for the anisotropy threshold.

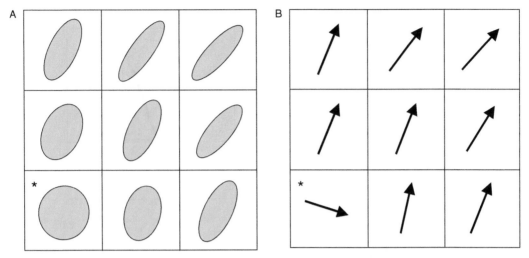

Fig. 9.2 Two types of fields we can use for tractography: tensor field (A) and vector field (B). We often use the vector field (B) for tractography, which represents the longest axis of the tensor (A). The vector may not be representing a fiber orientation when the tensor is isotropic or has a disk shape (the pixel with an asterisk).

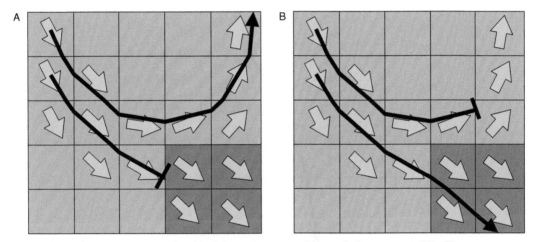

Fig. 9.3 Two types of termination criteria: low anisotropy (A) and sharp turn (B). Short arrows represent fiber orientations in each pixel and shading represents the degree of anisotropy.

9.4 SIMPLE STREAMLINE TRACKING CAN BE USED TO RECONSTRUCT A TRACT

The most intuitive and simplest way of reconstructing a 3D trajectory is to use a 3D vector field and propagate a tract linearly from a seed by following the local vector orientation. However, if a line is propagated simply by connecting pixels, which are discrete entities, the vector information contained at each pixel may not be fully reflected in the propagation. In the simple example illustrated in Fig. 9.4A, axonal tracts run at 30° from the vertical line. When applying the discrete "pixel connection" approach, a judgment has to be made about which pixel should be connected; for instance,

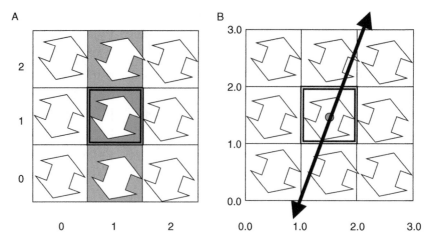

Fig. 9.4 Examples of simple linear tract propagation models. One of the most intuitive ways is to connect pixels based on fiber orientations (A). However, we encounter a resolution problem: namely, for a given pixel, we are obliged to pick one of the 26 surrounding pixels to connect (or 8 pixels in this 2D example) no matter what the fiber orientation is. This leads to inaccurate fiber tracing. In this example in (A), pixels are connected vertically, although measured fiber angles are slightly tilted to the clockwise orientation. The tracing accuracy can be improved by propagating a line continuously (B). To do this, we first have to convert the discrete (integer) pixel coordinates to continuous floating point coordinates as indicated in (B).

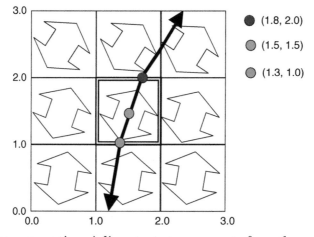

Fig. 9.5 Simple linear tract propagation. A line starts to propagate from the center of the seed pixel (the coordinate [1.5, 1.5]) based on the fiber orientation within the pixel. The line exits the seed pixel at coordinate [1.8, 2.0] and changes direction based on the second pixel.

is the 30° vector angle pointing at pixel {1, 2} or {2, 2}? No matter what the judgment is, it is clear that this simple pixel connection scheme cannot represent the real tract even in such a simple case. The simplest way to solve this issue is to convert the discrete voxel information into a continuous coordinate and linearly propagate a tract, in the continuous number field. This conversion from the discrete to continuous number field is shown in Fig. 9.4B.

This idea can be applied to a nonhomogeneous vector field, as shown in Fig. 9.5. In this example, tracking is initiated from the center of a pixel at coordinate (1.5, 1.5) in the continuous coordinate. A line is propagated in the continuous coordinate along the

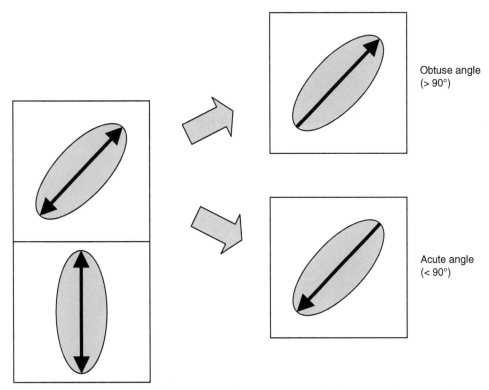

Fig. 9.6 There are always two options when two pixels are connected. One is an obtuse angle and the other an acute angle. A line is propagated by always choosing the obtuse angle. When fiber angles of two connected pixels form a 90° angle, we cannot decide which way the line should be propagated. Tract propagation should be terminated in such a situation.

direction of the vector of the pixel. Then the line exits the initiating pixel at the coordinate of, e.g., (1.8, 2.0) and enters the next pixel, in which the tracking starts to observe the vector direction of the new pixel.

The orientation of the longest axis of the diffusion ellipsoid does not have polarity. Therefore, we have to propagate the line to two opposite directions. In this case, the line exits the seed pixel at two points (1.8, 2.0) and (1.3, 1.0). There is also an interesting problem because of this nonpolarity issue. Every time a line exits from a pixel and enters the adjacent pixel, there are always two options for propagating the line (Fig. 9.6). We always choose the vector orientation that is pointing away from the incoming line (obtuse angle). What if the vectors of two connected pixels are crossed at 90°? Should we choose to go right or left? Apparently, the solution becomes very unstable, because 89° and 91° lead to two completely different pathways. Here, the angle threshold comes in handy. If the threshold is set at, for example, 45°, the tracking stops at such a problematic region. Also, the seed does not have to be at the center of a pixel or only one in each pixel. Seed density and locations within a pixel can be varied.

In this linear propagation approach, there is always a kink where the line exits from one pixel to the other. Smoother (curved)

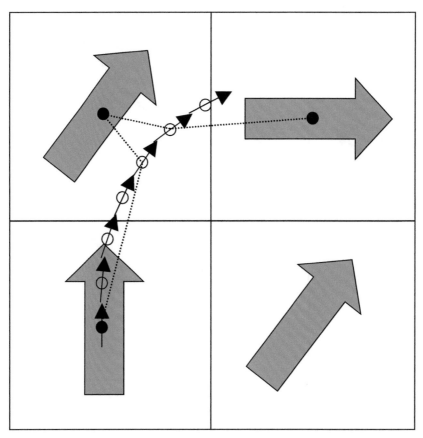

Fig. 9.7 An example for propagating a smooth curve based on a discrete vector field. For each small step with a fixed length, a new fiber orientation is calculated. In this example, the orientation is calculated from a weighted average of vectors in two nearest pixels. (Figure reproduced from Mori & van Zijl, *NMR Biomed*, 15, 468–480, 2002 with permission.)

paths can be created by using one of the interpolation methods, as shown in Fig. 9.7. In this example, a line is propagated with a small, predefined step size. Whenever it moves to a new coordinate, a distance-averaged vector orientation is calculated. In this most simple example, the average between two pixels closest to a new coordinate is used to draw a smooth line between pixels. By using this approach, a path could be traced more smoothly when there is a sharp angle transition. However, care must be taken because the tensor model itself assumes that fiber orientation is uniform and linear within a pixel, i.e., we have an assumption that our imaging resolution is high with respect to the curvature of white matter tracts, so that we can consider the tracts are linear within a pixel. It does not make much sense to use highly elaborate and time-consuming mathematics to draw curvature within a pixel, because the idea of drawing more accurate curvature within a pixel already violates the assumption of tensor calculation prior to the 3D fiber tracking. Probably, terminating the tracking when there is a sharp angle transition is the most conservative way to perform the tracking.

9.5 THERE ARE MANY LIMITATIONS TO SIMPLE TRACT PROPAGATION METHODS

9.5.1 Noise

Raw images in DTI contain noise and, as a consequence, the calculated vector direction may deviate from real fiber orientations. One of the drawbacks of tract propagation methods is that noise errors accumulate as the propagation becomes longer. The extent of these errors depends on the shape of the trajectory, anisotropy, resolution, and the interpolation method used.

9.5.2 Partial volume effect

Pixels that fall between two unrelated fibers could have a fiber angle that is the population-weighted average of the two fiber angles. Anatomic information that degenerates within a pixel is not easy to recover. Tractography results always contain false results due to the partial volume effect.

9.5.3 Branching

Axons from individual neurons often have branches. Axons merge into and exit from axonal bundles. In line propagation methods, there is only one line propagated from one seed pixel, as shown in Fig. 9.4, and branching cannot be properly represented.

9.5.4 Disk-shape anisotropy

When diffusion is isotropic, we can avoid performing tracking in such regions by using an anisotropy threshold. However, there are regions in the human brain where anisotropy is relatively high but has a disk shape (see Chapter 8). In such regions, the orientation of the longest axis becomes randomized, and, thus, it is likely that the propagation is terminated if an angle threshold is used (Fig. 9.3B). However, it is also possible that such a pixel causes a substantial error in tract propagation if it is not terminated. If the disk-shaped anisotropy results from crossing fibers, non-tensor methods may be able to resolve the two fibers.

9.5.5 Crossing fibers

Because the tensor model cannot appropriately represent pixels with multiple populations, all tracking techniques based on the tensor model may fail in brain regions with significant fiber crossing. There are two possible outcomes. One is false negative, in which tracking terminates in such regions. Fiber crossing regions tend to have low anisotropy and random fiber orientations, which lead to termination of fiber tracking. The other outcome is false positive.

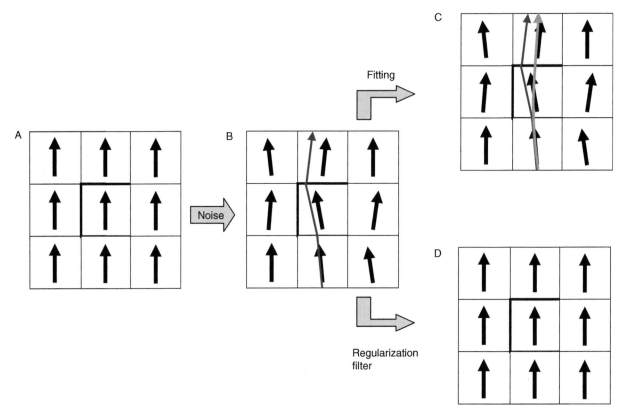

Fig. 9.8 Effect of regularization to reduce noise error. Suppose there is a uniform fiber structure as shown in (A). A vector field obtained from a DTI measurement may not correctly represent the fiber orientations due to noise, and tracking results are affected by the measurement error (red line in B). To reduce the effect of noise error, a fitting-based method can be used, in which vector information from more than two pixels is used to calculate a fiber path. This approach tends to smooth out the noise error if it is purely random (green line in C). Alternatively, we can denoise the vector field by using a regularization filter.

This can be further divided into two classes: (1) bias, in which tracking is shifted from the real path in a reproducible manner when it penetrates the problematic regions; and (2) switching, in which tracking switches from a tract of interest to an unrelated crossing tract.

9.6 SEVERAL APPROACHES ARE PROPOSED TO TACKLE THE LIMITATIONS

9.6.1 Smoothing using fitting or regularization

The noise effect can be reduced using smoothing techniques that average the vector or tensor information among neighboring pixels with a cost in the reduction in effective resolution. Several approaches to minimize noise effects have also been suggested, such as approximation of a tensor field based on B-spline fitting or "regularization" of the tensor field based on the "low-curvature hypothesis." The idea is illustrated in Fig. 9.8. Suppose there is a uniform vector field (Fig. 9.8A), which is disturbed by random noise (Fig. 9.8B). The simple tract propagation method faithfully follows the noise-affected vectors (red line in Fig. 9.8B).

Fig. 9.9 Examples of the tensor line approach. Blue lines indicate incoming fiber angles and brown lines outgoing fiber angles. (Figures were provided courtesy of Drs. Alexander and Lazar, University of Wisconsin.)

The fitting-based method incorporates information from at least two connected pixels and fits a smooth line (green line in Fig. 9.8C), in which the noise effect tends to be canceled out. In this fitting-based method, vector information is smoothed while a line is propagated. Alternatively, the tensor or vector field can be denoised by using a regularization filter (Fig. 9.8D) before tractography. In this approach, noise is reduced by integrating information of neighboring pixels. In other words, the vector orientation of each pixel is forced to be similar to those of its neighbors.

9.6.2   Tensor line approach

One of the approaches proposed to deal with disk-shape anisotropy is the "tensor-line" approach. This approach could be robust against noise. In this approach, the tracking tends to carry the momentum of the incoming pathways in lower-anisotropy regions (i.e., it tends

Vector Generation

DATA   #1   #2   #3   #4

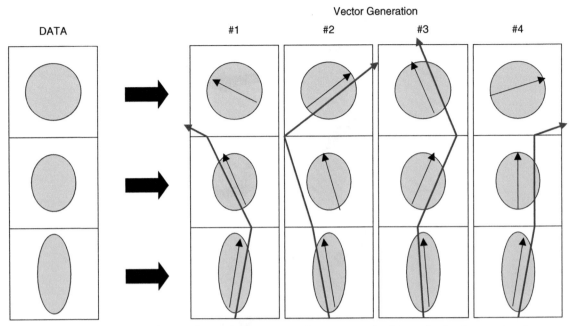

Fig. 9.10 Schematic diagram of a random vector generation approach. At each pixel, a vector is generated. The vector generation is random but its orientation is skewed, depending on the shape of the diffusion ellipsoid. When the shape is a sphere or a disk, the orientation is completely random (upper row). As the ellipsoid elongates, the vector orientations become more skewed to the longest axis (bottom row). After each vector generation, fiber tracking is performed. The results become more dispersed once the tracking hits a spherical or disk-shaped ellipsoid.

to linearly penetrate the low-anisotropy regions). The incoming line orientation ($\mathbf{v}_{in}$) is modulated according to the orientation of the diffusion ellipsoid, and the outgoing vector $\mathbf{v}_{out}$ is calculated according to the equation:

$$\mathbf{v}_{out} = \overline{\overline{\mathbf{D}}}\mathbf{v}_{in}$$

Figure 9.9 shows how this equation behaves in different situations. When $\overline{\overline{\mathbf{D}}}$ is spherical or the incoming line is parallel to the flat plane of the disk-shaped ellipsoid, $\mathbf{v}_{in} = \mathbf{v}_{out}$. In other words, the line extrapolates the incoming direction, $\mathbf{v}_{in}$, and passes through the pixel without changing its orientation. If the incoming line hits the planar-shape ellipsoid from an oblique angle, it is deflected closer to the plane of the flat surface of the ellipsoid.

9.6.3   Random vector generation approach

In this approach, dubbed RAVE (RAndom VEctor) perturbation or PICo (Probabilistic Index of Connectivity), a vector is generated at each pixel prior to tractography, and this process is repeated many times. As shown in Fig. 9.10, the generated vector orientation is completely random when there is no anisotropy (or disk shape). The orientation probability is skewed to the longest axis in anisotropic regions; the higher the anisotropy, the more skewed the vector orientation. After many repetitions, bundles of tracking

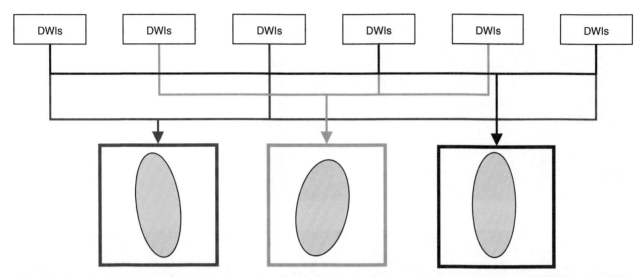

Fig. 9.11 Schematic diagram of a bootstrap method. Suppose there are six repeated sets of DWIs (each DWI dataset contains complete data for DTI), a tensor field is calculated using a subset of the six datasets. In this case, three DWI datasets are randomly chosen to calculate a tensor field. We can repeat this procedure many times using different combinations of the six datasets. The resultant tensor fields may not be completely the same due to noise.

results are generated, which carries information on the reliability range of the tracking results. This approach could be useful in identifying problematic regions where anisotropy is low or disk shaped and the vector of the longest axis has low reliability. For details of this approach, see the papers by Lazer et al. (RAVE) and Parker et al. (PICo) in the reference list at the end of this chapter.

9.6.4 Bootstrap approach

In the random vector generation approach described in the preceding subsection, it is assumed that the shape of the diffusion ellipsoid carries the probability of the fiber orientation. However, the diffusion ellipsoid itself could contain error by noise. In the bootstrap approach, the tensor field is generated repeatedly using different sets of raw DWIs (Fig. 9.11). In this example, six complete sets of DWI (e.g., six repeated datasets of six-orientation measurements) are acquired, and three of them are used to calculate a tensor field. The combination of the three DWIs is randomly decided. By repeating this process, many tensor fields are obtained. Using one of the tracking methods, a tracking result for each combination can be generated, which carries information about the reliability range of reconstructed trajectories, as shown in Fig. 9.12. This is an excellent approach to investigate the effect of noise and provide reliability information about reconstructed trajectories. However, it requires rather large datasets of DWIs, and its applicability to clinical studies is yet to be seen.

A

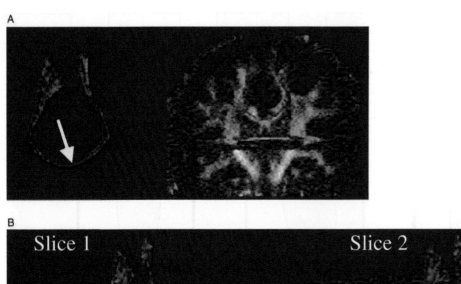

B

C

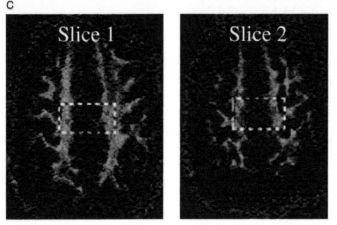

Fig. 9.12 An example of the bootstrap result. A seed was placed in the mid-sagittal region of the corpus callosum (yellow arrow in (a)). After a large number of bootstrap trials, a bundle of tracts were generated from the single-pixel seed. The corresponding fiber densities are displayed for two axial slice locations in (b: 3D) and (c: 2D slices). (The figures are reproduced from Lazar and Alexander, *Neuroimage*, 24, 524–532, 2005, with permission.)

9.6.5   First marching approach

In this approach, a three-dimensional surface is propagated from a seed pixel. The propagation rate of the front line is determined by the tensor at each pixel. The rate is highest along the longest axis and slowest along the shortest axis. This is similar to the way stain propagates on a sheet of cloth. In isotropic regions, the stain propagates in all directions. By capturing the shape of the propagating front line at a fixed time interval, a map that is similar to geological maps can be obtained. Using this analogy, mountain regions (parallel to the longest axis of diffusion ellipsoids) have slow propagation, where the front lines captured at the fixed time interval are close together. Along the longest axis, the intervals of the front lines are wide, forming a valley. Once a 3D map is completed, we can consider it as a connectivity map; the valley regions have a higher probability of connectivity. This is not a deterministic approach; it can follow branching regions. The degree of connectivity probability can be represented by the time it takes to reach the area of interest. The most probable path toward the seeding pixel can be reconstructed from an arbitrary point, in the same way that water can find a river. This technique could also penetrate problematic regions with crossing fibers or be combined with non-tensor models. For details on this approach, see the papers by Parker et al. (2002) in the reference list at the end of this chapter.

9.6.6 Simulating annealing and spaghetti models

If two anatomical regions are known to be connected, the most probable path could be searched by an energy minimization approach. In this approach, there are usually two competing terms in the equation, and we search for a condition that can minimize both terms. One of the terms is alignment with measured diffusion tensor. We need to devise a cost function that becomes smaller if a path is more aligned to the longest axis of the tensor. The other competing term is usually the simplicity of the trajectory such as the length of the path and the amount of curvature. The simplest way to connect two points is to use a straight line (thus, the second term is minimized). However, such a solution is unfavorable with respect to the first term, because this straight line is most likely not aligned with the tensor field. If the pathway is allowed to be extremely flexible, there might be numerous solutions that could connect the points with good alignment with the tensors. A path generated by this technique is based on the balance between these two terms. With an appropriate amount of allowed stiffness, the path should be able to find a way to reproduce the correct trajectories. The best pathway is searched using, for example, simulated annealing. This technique is an attractive choice for dealing with non-tensor models, in which there are often several fiber populations within a pixel and tractography needs to penetrate problematic regions (where fibers are crossed). For details about this approach, see the papers by Parker et al. and Tuch et al. (2000) in the reference list at the end of this chapter.

9.6.7 Differences of various tractography approaches

One of the most difficult questions is "which tracking algorithm gives the right answer" or "which tracking algorithm should be used." There are two ways to answer this question. First, if we are

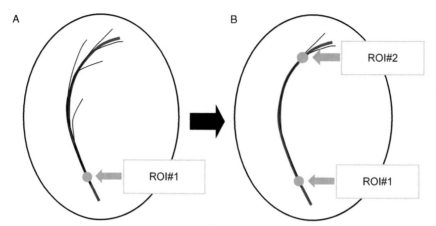

Fig. 9.13 Schematic diagram of a multiple-ROI approach. The red line represents the real path of a white matter tract. If the first ROI is five pixels large, there would be five tracking lines, some of which may deviate from the path. The deviated path may represent a real branch but could be errors due to noise and partial volume effect. If the destination of the tract is known, the second ROI can be placed, which selects tracking results that penetrate both ROIs.

interested in core regions of prominent fibers, all algorithms should provide similar results. We could limit ourselves to studying these regions of white matter. If one needs to study white matter outside the core regions, this approach may not be satisfactory. Alternatively, we can use the same algorithm for the entire study and compare the results from different groups statistically. We can consider tractography as a tool for systematically extracting pixels that belong to a specific tract system. In this respect, different tractography methods have different ways of grouping pixels. This is not a matter of right or wrong: as long as the method provides reproducible results and the same method is applied to all data, the results should be meaningful and comparable. The interpretation of the results, however, does require great care. Quantification and reproducibility measurements are of great importance too (see the following chapter).

9.7   TRACT EDITING USES MULTIPLE REGIONS OF INTEREST

One of the most effective ways of dealing with errors by noise, partial volume effect, and crossing fibers is tract editing based on *a priori* anatomical knowledge. This technique can be considered as a method to remove false positives. Suppose the red line in Fig. 9.13A is the real trajectory of a tract of interest, and the first region of interest (ROI) is placed on the tract; if the ROI is five pixels large, there are five lines propagated from each pixel. Some of the tracking results may deviate from the real path due to noise, partial volume effect, or they could be real branches. Usually, we cannot judge which results are real and which are not (false positive). If there is *a priori* knowledge about the destination of the tract, we can draw the second ROI and retrieve

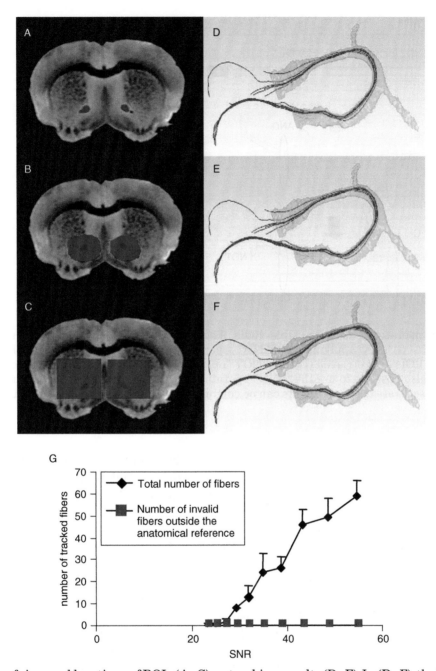

Fig. 9.14 Effects of sizes and locations of ROIs (A–C) on tracking results (D–F). In (D–F), the blue structure is the anterior commissure defined by a T_2-weighted image, and red lines are tracking results. The graph in (G) shows the number of fibers that penetrate the two ROIs as a function of SNR. The red squares show the number of false positives. (Images reproduced from Huang et al., *Magn. Reson. Med.* 52, 559–565, 2004, with permission.)

tracking results that penetrate both ROIs. The idea is that if tracking deviates from a real path, it is highly unlikely that it comes back to the real path in 3D space by chance and penetrates the second ROI.

Figure 9.14 shows experiments of the two-ROI approach using a high-SNR dataset from a postmortem mouse brain. The anterior commissure is a simple and well-isolated tubular white matter

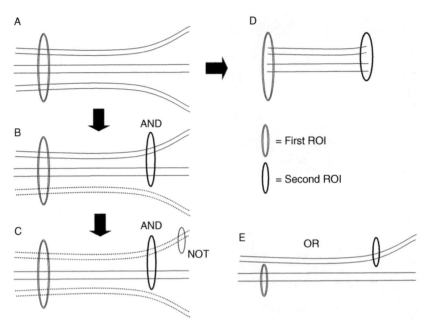

Fig. 9.15 Various operations with multiple ROIs. The one-ROI approach generates anatomically less constrained results (A). By applying an "AND" operation to the second ROI, tracking results that penetrate the two ROIs are retrieved (B). If a "NOT" operation is applied, fibers that penetrate the ROI are removed (C). It is also possible to use the "AND" operation to retain only the tract coordinates between the two ROIs (D). Two separate tracking results can be combined by an "OR" operation (E). Red lines are selected fibers.

tract that can easily be identified using T_2-weighted images. The trajectory was reconstructed from a 3D T_2-weighted image manually, which served as a gold standard. Locations and shapes of two ROIs are shown in Fig. 9.14a–c. As long as the ROIs are drawn with anatomically sound protocols, the results are rather insensitive to the way they are drawn. In this case, unless the corpus callosum is included in the two ROIs, the anterior commissure is the only tract that can satisfy the ROIs drawn in the two separate hemispheres. If noise is added to the raw data and the number of false results (deviation from the real path, yet penetrating the two ROIs) counted, the number of false results is always zero, but the number of reconstructed fibers that penetrate the two ROIs decreases as the noise increases (higher probability of false negative). Of course, this is a result of one tract, and we cannot generalize the result and say that "tracking results that penetrate two ROIs are always correct," but the result demonstrates the robustness of the two-ROI approach against false results and ROI drawing.

In the preceding example, two ROIs are connected by an "AND" operation (Fig. 9.15B), i.e., tracking results that penetrate the first AND second ROI are searched. There is also an option that removes portions of the reconstructed tract outside the two ROIs (Fig. 9.15D). These outside regions are not constrained by the two ROIs, and their reliability is different from the region between the two ROIs. If one is interested in tracking-based quantification, this operation may increase reproducibility.

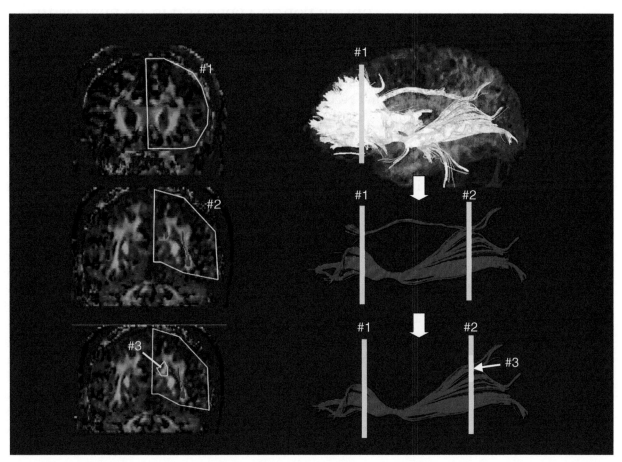

Fig. 9.16 An example of fiber editing using the "AND" and "NOT" operation to reconstruct the inferior fronto-occipital fasciculus (IFO). First, a large ROI (ROI #1) delineating a large portion of the frontal lobe is drawn using a coronal slice. This leads to a large number of trajectories that penetrate the ROI. The second ROI (ROI #2) is drawn at the occipital lobe, and an "AND" operation is used to search for only trajectories that penetrate both ROIs. This leads to delineation of the IFO. However, a part of the cingulum also penetrates the same ROIs. This cingulum component is removed by the third ROI (ROI #3) with a "NOT" operation (yellow ROI). (Images are reproduced from *MRI Atlas of Human White Matter*, Elsevier.)

If there is contamination that shares the two ROIs with the tract of interest, such contamination could be removed by a "NOT" operation (Fig. 9.15C). Two separate tracking results could be combined by an "OR" operation (Fig. 9.15E).

In Fig. 9.16, an actual process of fiber editing is shown. The tract of interest is the inferior fronto-occipital fasciculus (IFO), which has been described by neuroanatomists as connecting the frontal and occipital lobes. The first ROI is placed on the entire left frontal lobe, which leads to construction of numerous tracts. The second ROI delineates the occipital lobe. Upon inspection, there are only two tract families that satisfy these two ROIs: one is the IFO and the other is the cingulum, which is contamination in this case. The third ROI is placed on the cingulum, and a "NOT" operation is used to remove it. Using this simple three-step protocol, the IFO, which agrees very well with classical anatomical definition, can be obtained.

9.8   BRUTE-FORCE APPROACH IS AN EFFECTIVE TECHNIQUE FOR COMPREHENSIVE TRACT RECONSTRUCTION

Brute-force approach is a technique that can reduce the effect of noise and reconstruct branching/merging tracts. The idea is very simple. Usually, fiber tracking starts after a group of pixels are specified as an ROI, and these pixels serve as seeds. If the ROI is five pixels large and the seed density is 1 seed/pixel, there would be five lines propagated from the ROI. The number of lines increases by increasing the seed density, but tracking may not fully delineate the tract of interest if there are diverging branches.

Figure 9.17 illustrates the fiber reconstruction process. Let us suppose that there is a tract with the structure shown in Fig 9.17A. DTI measurement provides the fiber orientation information shown in Fig. 9.17B. For fiber reconstruction, a tract of interest needs to be identified first and marked as an ROI (bold box). Two possible reconstruction approaches are demonstrated. In the first so-called "from-ROI" approach (Fig. 9.17C), tracking originates from the ROI. In this simple example, the ROI contains only one pixel and, as a result, only one line is produced in the "from-ROI" approach, which would delineate only a part of the tract. In the second, so-called "brute-force" approach, tracking is initiated from all pixels, and all tracts that penetrate the ROI are retrieved. This approach is much more time consuming but leads to a more comprehensive delineation of the tracts.

9.9   ACCURACY AND PRECISION ARE IMPORTANT FACTORS TO BE CONSIDERED

To use tractography for application studies, it is imperative to know its limitations with respect to accuracy and precision. Accuracy means how real the result is (validity), and precision means how reproducible the measurement is (Fig. 9.18). A frequent criticism of tractography is that its accuracy and precision are not well characterized. It is very important to realize that both accuracy and precision depend on the biological questions we are trying to answer and how we use the tracking results. Even the same tracking results may lead to different biological conclusions depending on how we use and interpret the data, and the degree of accuracy and precision may vary (Fig. 9.19). For example, if we want to use tractography to segment the corticospinal tract in the medulla—lower pons level, the result could be both accurate and precise. Even without tractography, the tract can be readily identified in color maps. The tracking result agrees with classical anatomical description (thus validated), and the trajectory is well compacted and linear. By using a simple tracking protocol, we can reproducibly reconstruct the corticospinal tract.

On the other hand, if one wants to segment the corticospinal tract in the upper pons—motor cortex regions, the result may have unknown accuracy. In this region, the corticospinal tract runs

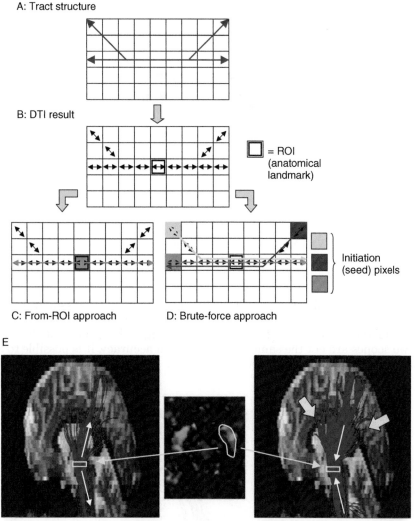

Fig. 9.17 Principles of tract reconstruction using the "from ROI" and the "brute-force" approaches. (A) Example of a tract structure with two branching points. (B) Results of DTI measurement; a vector field that shows the fiber orientation at each pixel. The bold box shows the anatomical landmark where a ROI is defined. (C) Results of the tract reconstruction using the "from ROI" approach, in which tracking begins from the ROI. This generally leads to an incomplete delineation of the tract. (D) Results of the "brute-force" approach, in which tracking begins from all pixels. Several initiation (seed) pixels from which the tracking can lead to the same ROI are demonstrated. (E) Results of actual tracking using the cerebral peduncle as a reference ROI. The left and right panels show results for the "from ROI" and "brute-force" approaches, respectively. (Images reproduced from *MRI Atlas of Human White Matter*, Elsevier.)

with many other corticoefferent fibers, and its location is debated even among neuroanatomists. Therefore, we do not have a means of validating our tracking results (thus, there is unknown accuracy). However, with an appropriate reconstruction protocol, we can reconstruct the trajectory with high reproducibility. We can use the trajectory information to segment pixels that may belong to the corticospinal tract and measure the size or FA of the segmented region. It carries trajectory information that may be used to identify involvement of stroke infarction or displacement by tumor growth, for example. We can take the same reconstruction data to study pixel-to-pixel connectivity between the cortex and the pons, the cerebellum, or the spinal cords. Validity of these point-to-point

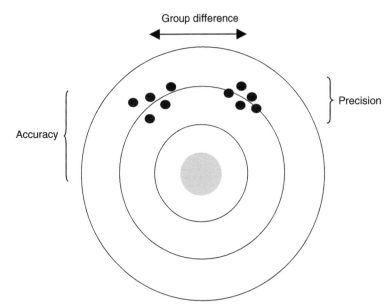

Fig. 9.18 Concept of accuracy, precision, and group difference. Suppose a marksman shoots a bull's eye. If the shots are well clustered, the marksman (or the rifle) is precise. However, in this case, his scope is not well adjusted, and the shots are not accurate. The marksman changes the scope and shoots again. The results indicate that the two scopes are not the same. Without known accuracy, it is possible to study group difference, although care must be taken in interpreting the result.

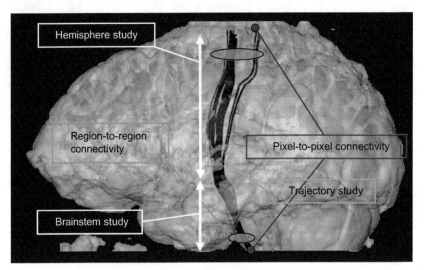

Fig. 9.19 Different ways to use the same tracking results. If one is interested in connectivity, one can study which pixels are connected (pixel-to-pixel connectivity) or which regions are connected (region-to-region connectivity). The former information may have lower accuracy and precision. If one is interested in how the tract is deformed by tumor growth, the tract trajectory is what he/she wants to know. Its accuracy and precision may not be the same for the brainstem and cerebral hemisphere.

connectivity results is even harder to establish, and the reproducibility could be poor because the results are extremely sensitive to noise, partial volume effect, and image resolution.

The preceding examples show that the accuracy and precision of tractography depend critically on the biological questions and the way tractography is used. In other words, we cannot provide

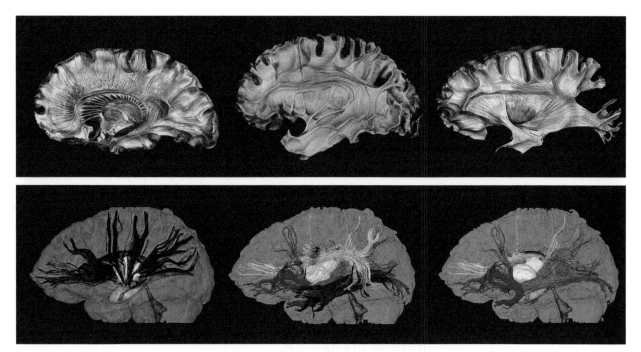

Fig. 9.20 Comparison between postmortem- (upper row) and DTI-based reconstruction results (bottom row). (Copyright-protected materials (postmortem tissue images). The postmorterm tissue images are reproduced from Atlas Cerebri Humani, S. Kager AG, Basel with permission.

generalized answers about accuracy and precision unless we know our biological questions.

9.10 REPRODUCIBILITY OF TRACTOGRAPHY IS MEASURABLE

Reproducibility (precision) of tractography is measurable. Two major sources of variability are data acquisition (e.g., noise, partial volume effect, and physiological motion) and ROI placement. The reproducibility of data acquisition can be measured by scanning the same person multiple times, aligning the data in 3D, and calculating standard deviations of anisotropy in each pixel. This obviously depends on SNR, scanning times, and pixel resolution. The degree to which the subjects cooperate is also a large factor. The sensitivity to ROI sizes and locations decreases by using two ROIs with a sensible protocol for their placement. By repeating the reconstruction of the same tract using the same dataset by the same operator (intra-operator) and by different operators (inter-operator), reproducibility can be measured. However, reproducibility is not the same for different tracts and depends on their size and trajectory. Therefore, it is not possible to generalize the reproducibility of tractography. For each tract of interest and acquisition protocol, the reproducibility needs to be measured.

In our experience, reproducibility of DTI data acquisition and tractography is relatively high if one is interested in the core regions of major white matter tracts. For example, with a 1.5T magnet, 2.5 mm isotropic resolution, and a total of 90 DWIs (approximately 15 min scans), the average standard deviation of FA values is approximately within 3% for the white matter (FA > 0.2), calculated from three repeated measurements of the same subject.

While the above-mentioned measurements may indicate reasonable reproducibility of DTI and tractography, it does not necessarily mean that tractography is a sensitive quantitative tool for detecting abnormalities in the white matter, because there are several important factors that we have not yet discussed. First, tractography results tend to have a large amount of variation among normal subjects. For example, if the same tractography protocol is applied to two normal subjects and the numbers of pixels (i.e., the size of the reconstructed tract) are compared, it is not uncommon to find a 30% difference. This large variation is caused by the fact that tractography integrates all anatomical differences along the tract. This point is discussed later in this chapter. If the range of normal deviation is large, the sensitivity to detect abnormality becomes low. Second, we need quantification tools to compare results from different subjects or different groups. Analyses of shapes and volumes are in general a challenging task. This issue will be discussed in Chapter 10.

9.11   TRACTOGRAPHY REVEALS MACROSCOPIC WHITE MATTER ANATOMY

An actual process of tract reconstruction is shown in Fig. 9.16. The process contains a subjective judgment in ROI placement, and the results depend on parameters such as thresholds, ROI sizes, and locations. However, some tracts of interest can be obtained reproducibly with a relatively simple protocol. In Fig. 9.20, the reconstruction results of several prominent white matter tracts are compared with postmortem specimens. In these figures, major fiber bundles are reconstructed as faithfully as possible to the known tract trajectories using multiple ROIs. Macroscopically, the agreement is excellent. Excellent agreements are also reported in animal studies where tracking results are compared with existing atlas or manganese-based tract-tracing studies. These results suggest that tractography has the ability to detect real tract pathways. However, these types of macroscopic agreement do not necessarily mean that the results are valid at a microscopic level, because each propagated line does not correspond to actual axons or brain connectivity. In DTI, what we are observing is the motion of water molecules, and the imaging resolution is on the order of 1–5 mm. It is not possible to reconstruct individual axons whose diameter is 1–10 μm. What we can see from the postmortem samples in Fig. 9.20 are bundles of axons. Actual axons merge and branch from the bundles at various points. Even if we could accurately follow the path of large bundles, that does not mean we can study individual axonal paths or connectivity of axons.

When we discuss validity, it is important to define what "connectivity" is. It is tempting to think that tracking results reveal

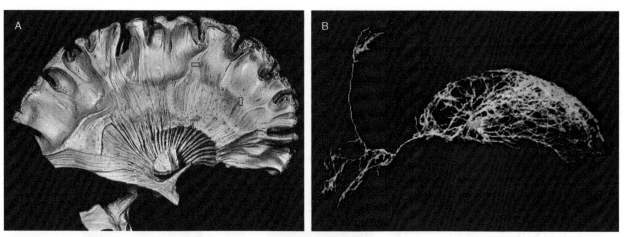

Fig. 9.21 Comparison of a postmortem human sample that reveals the macroscopic white matter architecture (A) and single-cell reconstruction of a neuron in a rat brain using an invasive *in vivo* chemical tracer experiment (B). In (A), the approximate size of 1×3 mm pixel is shown by red boxes. The image in (A) was reproduced from Atlas Cerebri Humani, S. Kager AG, Basel with permission and in (B) was reproduced from *The Human Brain*, 1988, Mosby, with permission.

the cell-level neural connectivity, but what is seen in Fig. 9.20 is not axonal connectivity, but the macroscopic anatomy of the white matter. In Fig. 9.21A, 1 × 3 mm boxes are placed on a postmortem specimen. This size of probe could delineate overall configuration of the white matter anatomy seen in Fig. 9.21A. In Fig. 9.21B, the result of single-neuron reconstruction in a rat hippocampus is shown. There are two important facts in this figure. First, with the current level of image resolution, it is impossible to reveal the single-cell level connectivity. Second, even if we could reconstruct the entire neuron, "connectivity" is not as simple as connecting two points. Neurons have a dendrite network to communicate with nearby neurons. They have axons to communicate with distant neurons, which could have many branches. If biological questions require information about the cell-to-cell-level connectivity, DTI and tractography may not be the right tool. The real power of DTI is to delineate macroscopic white matter organization in roughly 10 min of scanning time, which cannot be achieved by invasive chemical tracer technologies. There are many important biological and clinical questions that can be answered by DTI but not by microscopic methods. We need to use the right tools for right questions.

9.12 THERE ARE ROUGHLY THREE TYPES OF INFORMATION OBTAINED FROM TRACTOGRAPHY

Tractography provides one of the three types of anatomic information. First, it provides connectivity information. As shown in Fig. 9.22A, the start and end points of tracking become important information in this case. Second, tractography can be considered as a tool to semiautomatically group pixels (or a region-growing

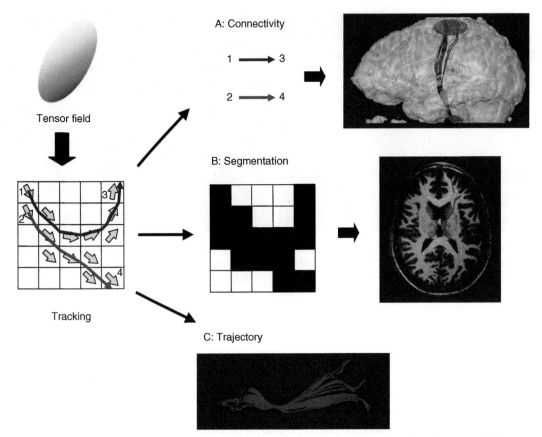

Fig. 9.22 Three ways to use tracking results: segmentation, connectivity, and trajectory. When a tracking result is used for white matter segmentation, its information is usually converted to pixel-by-pixel binary information, which indicates pixels occupied by the reconstructed tracts. When connectivity is studied, the start and end points of the tract are of interest.

segmentation tool) based on fiber orientation (Fig. 9.22B). When tractography is used for segmentation, its information is usually converted to pixel-by-pixel binary information. To study connectivity, it is necessary to follow the start and end points of each streamline. Once the streamline information is converted to binary information, specific connectivity information is lost because we can no longer tell which pixel is connected to which. Using the binary information, the size or various MR properties of the segmented area can be quantified by superimposing it on any coregistered MRI map. The third information that can be obtained from tractography is tract trajectories (Fig. 9.22C). This could be important information if one wants to delineate altered white matter anatomy due to brain tumor or developmental abnormality. Tractography can be used to identify locations of tracts of interest in such cases.

Figure 9.23 shows an interesting extension of the tractography-based segmentation approach, in which the cerebral peduncle is segmented into three areas based on their projection to the frontal lobe, motor cortex, and parietal/occipital lobes. These three areas agree with known anatomy (Fig. 9.23B) and cannot be distinguished by color-coded maps. This type of segmentation is

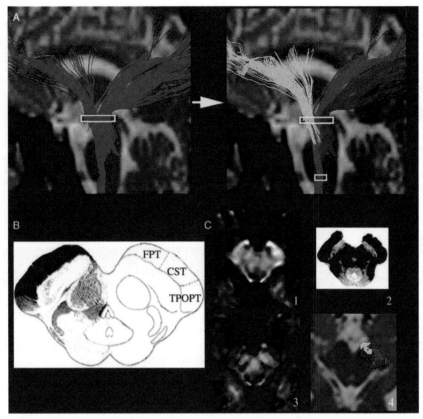

Fig. 9.23 Segmentation of the cerebral peduncle based on tracking. The cerebral peduncle is known to contain three major white matter tracts with different connectivity (B). These are frontopontine tract (FTP), corticospinal tract (CST), and temporo-/parieto-/occipitopontine tract (TPOPT). However, these tracts run parallel in the cerebral peduncle and cannot be distinguished by anisotropy map (C-1), histology (C-2), and orientation map (C-3). However, it can be segmented based on trajectories of the fibers (C-4). In this example, a tracking result from the cerebral peduncle is divided into three families based on their trajectories (A). In this way, the cerebral peduncle can be segmented by incorporating connectivity information. The figure is reproduced from Stieltjes et al., *Neuroimage*, 14, 723, 2001 and the drawing in (B) was modified from *Carpenter's Human Neuroanatomy*, 1976, Williams & Wilkins.

possible only by incorporating tract trajectory and connectivity information distal to the cerebral peduncle. In Fig. 9.24, other examples of tractography-based segmentation are shown, in which thalamus and the entire white matter are parcellated based on connectivity to the cortex.

9.13   HOW CAN WE VALIDATE TRACTOGRAPHY?

Accuracy (validity) of tracking results has been a central issue in the DTI research field. From the preceding discussion, it is clear that a question such as "is the tracking result real?" is rather naïve. Researchers could draw different biological conclusions from the same tracking results, and we cannot validate the results without knowing the conclusion (Fig. 9.19).

An individual tracking result of *in vivo* human data could not be validated because the gold standard requires invasive observation such as chemical tracer experiments. Such studies can be

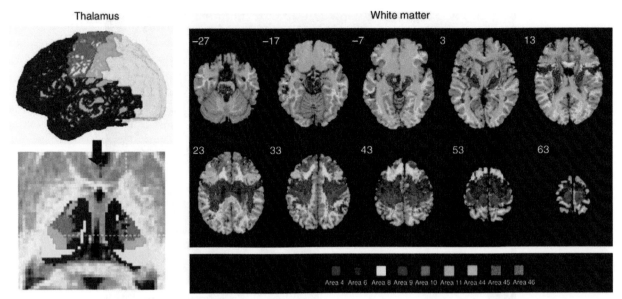

Fig. 9.24 Examples of structural parcellatoin based on tractography. In this type of studies, the cortex is parcellated first and the results extrapolated to other regions of brain, such as the thalamus and the white matter, using tractography. Images in the left column are provided coutesy of Drs. T.E. Behrens and H. Johansen-Berg, University of Oxford (Behrens et al., 2003) and in the right column are reproduced from Thottakara et al., *Neuroimage*, 29, 868, 2006 with permission.

performed using animal models, but whatever the findings are, we cannot extrapolate the result and declare that tracking is always true or always wrong in human studies. We know that validity of tracking results may vary if, for example, image resolution is different. In a sense, this is a common problem of MRI. Even gray matter–white matter contrast in T_1- and T_2-weighted images is difficult to validate. The gray matter volume could be different depending on T_1 or T_2, echo time, flip angle, magnetic field strength, or even the quality of the RF coil. However, we know that gray matter–white matter contrast is very useful to study brain anatomy. Clearly, the quest for accuracy should, at some point, be balanced by precision, usefulness, and practicality.

This is not to say that study for validity is not important. The first line of validation comes at the data acquisition stage, i.e., accuracy of pulse programming, b-value calculation, scanner stability, and gradient performance/calibration are important factors. At the level of tensor calculation, calculation errors may occur. These can be validated by using a phantom with known fiber orientations. For example, fiber orientation of a piece of meat can be measured before and after known rotation angles. At the level of fiber tracking, there may be a bug in the program. This can be checked using simulation data. All these processes can be validated by imaging fixed nervous tissues. Some isolated and tubular tissues such as the anterior commissure or optic nerves could be ideal systems. We can confirm that calculated vector angles are along the tract, and tracking can reconstruct the entire length of the pathway. For this type of simple nervous structures, no matter which tracking algorithm is used, we should get the same result.

After these validation processes, we can say that our DTI measurement and tracking software are accurate. This type of validation is absolutely necessary but there still remain two "validation barriers" that cannot be addressed from these types of validation studies. The first barrier is that, as mentioned above, we cannot generalize the validation results. No matter how accurate (or inaccurate) our tractography results are in a model system, we cannot conclude that tractography is always right (or always wrong) because we know that the results are sensitive to image resolution, noise, partial volume effect, the size of the tracts of interest, and so on. Validity could be different, depending on whether we are studying known fibers using the existing knowledge as anatomical constraints or trajectories of unknown tracts. Second, validity may vary depending on the final conclusion of the study. As described in the previous section, there are many ways of interpreting tracking results to study different types of biological questions (Fig. 9.19). Different researchers may draw different biological conclusions from the same tractography results. We cannot conclude validity solely based on a tracking result itself.

Another important fact is that the anatomical entity we would like to reconstruct is sometimes not well defined. For example, there are bundles of fibers called superior and inferior longitudinal fasciculus, superior and inferior fronto-occipital fasciculus, or corona radiata. These tracts can be identified in postmortem specimens but they are not precisely defined in terms of trajectory, boundary, connectivity, and functions. These tracts most likely consist of axonal tracts merging and branching at many points along the trajectory. Therefore, they may not represent specific functions or connectivity. Without clear definition and boundary, the validation of tracking results of these "collective" bundles could be an elusive target.

9.14   HOW SHOULD WE USE A TOOL WITH UNKNOWN ACCURACY?

Tractography is a tool with unknown accuracy in many cases. However, so are many tools in imaging research. As described earlier, validity of something as simple as T_1-based brain segmentation is not fully characterized. We can consider imaging as a tool to systematically reduce the vast anatomical information of the brain into merely 1 MB (T_1-weighted images) to 10 MB (DTI) of information. On the other hand, even the 10 MB of tensor information is often too large to fully exploit its anatomic content. Tractography is a tool to further reduce the information to typically much less than 1 MB. After this degree of information contraction, anatomic information may not represent a specific brain structure. Namely, in the 1–10 MB of information, a considerable amount of anatomic information degenerates, and a variety of anatomical configurations may result in the same MR results (see, e.g., Fig. 7.3). Therefore, it is often difficult to unequivocally determine the meaning of MR observation. In this respect, we can consider MRI as a means of systemically reducing the amount of anatomic information to a manageable size, allowing us to efficiently screen the status of the entire brain. This is a particularly powerful tool for comparing different subjects or groups (controls and patients) (Fig. 9.18).

If there are differences between two groups in tractography, we need to know if the differences are real and how to interpret them. For example, suppose we have the following hypothesis: the superior longitudinal fasciculus (SLF) of schizophrenia patients is smaller than that of normal subjects. We can establish a protocol to reproducibly reconstruct the SLF, measure its size, or evaluate connectivity, and we may find that the results are in line with the hypothesis. The important question is whether the tracking results truly reflect the SLF. We should not forget that what we measure is water diffusion, and the tensor model is based on bold simplifications. The advantage of imaging is that it can screen the entire brain in a systematic way. Imaging can quantitatively compare different brain regions or different subject groups to efficiently bring our attention to a subset of the brain structure. Tractography "indicates" a possible link between anatomy and functional outcomes. This type of observation should be followed up by other means such as manual ROI-based quantification or voxel-based morphometry. The results can also be combined with other modalities such as histology, fMRI, PET, or EEG. Conversely, if an empirical but robust link between tractography and a disease is established (e.g., the SLF is always small in a subtype of schizophrenia), it would become an important clinical marker. Either way, the most important issue is to clearly know what the advantages and disadvantages of tractography are, what the final goal of the study is, and which anatomic information we need from tractography to achieve the goal, and to design the study based on these considerations.

9.15   QUANTIFICATION IS A KEY TO MANY TYPES OF TRACTOGRAPHY-BASED STUDIES

As will be discussed in Chapter 11 (DTI Applications), there are two types of application studies. In the first category, tractography will be used for qualitative study for individual cases. Most routine radiological diagnoses are made in a qualitative manner. For this type of application study, abnormalities found by tractography need to be much larger than the extent of reproducibility errors and variations among normal subjects. The applications may also be limited to those tracts that have been well characterized by classical anatomic studies. The second category requires quantification. There are several cases in which quantification becomes critically important.

9.15.1 Only a small amount of abnormality is expected

In many geriatric and psychological studies, we do not expect large anatomical changes. In this type of study, the precision of tools needs to be well characterized. Even if we could design a reconstruction protocol with high precision, individual tracking results may contain errors due to noise and partial volume effects. There are variations in white matter anatomy within normal populations. Therefore, the average and the standard deviation of each group need to be calculated and statistically compared.

9.15.2 Unknown tracts are studied

One way of minimizing the problem of unknown accuracy is to study well-characterized white matter tracts. We could design multi-ROI protocols to reconstruct fibers as faithfully as possible to the known trajectory. Identification of known tracts may not have a scientifically high impact, but there should be many application studies that could benefit from the fact that we can delineate these tracts noninvasively. When we need to investigate previously nondescribed tracts, the unknown validity (i.e., does the tract really exist?) becomes the central issue. One way of addressing this issue is to investigate the probability of the findings. If a previously nondescribed tract is reproducibly found in all subjects, it is likely that the result indicates the existence of a coherent anatomical structure.

9.15.3 Anatomy–function correlation is studied

To perform correlation studies between anatomy and other parameters such as functional measurements, the anatomical properties of the tracts of interest (location, size, and shape) need to be quantified.

9.16 THERE ARE SEVERAL POSSIBLE REASONS THAT LEAD TO SMALLER (OR LARGER) RECONSTRUCTION RESULTS

In previous sections, it has been mentioned repeatedly that the interpretation of tractography results needs great care. If one uses tractography as a segmentation tool, we can measure the size of the tract of interest by simply counting the number of pixels that are occupied by the reconstructed tracts. The number of traced lines is also often used as an indicator of the size of the tract. An often-asked question is: what does the number of traced lines mean? Confusion may arise if one wants to connect the meaning of the traced lines to axons. One is a macroscopic entity and the other is microscopic; there is only a poor link between the two. The number of traced lines or the number of pixels occupied by the

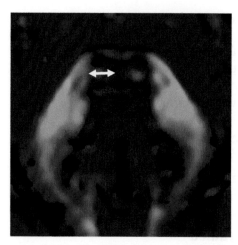

Fig. 9.25 In this color map at the pons, the corticospinal tract can be clearly identified, and its size is measurable. This is a cross-sectional view of the corticospinal tract, in which there are a large number of axons, running parallel. Its size may have a relationship to the number of axons, but the size change could be due to fewer axons or smaller diameter of each axon. In addition, this blue fiber bundle also contains the cortico-pontine and corticobulbar tracts.

lines are both indicators of the size of large fiber bundles which are collections of a huge number of axons that may merge and branch at any level of the bundles. Many axonal bundles have been documented by neuroanatomists and are clearly seen in color maps (Fig. 9.25). It is possible to measure the size of prominent bundles by using slice-by-slice inspection of color maps. However, such manual methods could be difficult if the tract changes its direction within the observation plane. The manual drawing process is time-consuming and possibly inaccurate. When one is interested in the tract size, tractography can be considered as a tool to assist the segmentation of the bundle, as explained in Fig. 9.22.

When the number of traced lines is chosen as an indicator of the size, care must be taken because the number does not have an anatomical meaning. The number could be doubled by increasing the density of seed (e.g., one seed/pixel *versus* two seeds/pixel). The pixel count approach may be the more straightforward way because it gives absolute volumes and should not be very sensitive to the seed density as long as the brute-force method is used. However, if a difference in a tract size is observed between groups, interpretation is not straightforward.

In Fig. 9.26, several reasons that could lead to a narrower reconstruction are listed; namely, narrowing of the tract, partial narrowing, perturbation by crossing fibers, and lower anisotropy. This is an example of how anatomic information degenerates: different anatomical structures lead to the same results, or in other words, the reconstructed tract does not necessarily represent the anatomy of the tract of interest. In some cases, the reason for the smaller tract size may be the anatomic changes in unrelated tracts (Fig. 9.26C). It is also important to point out that, as shown in Fig. 9.14, the size of the reconstructed tract is sensitive to the

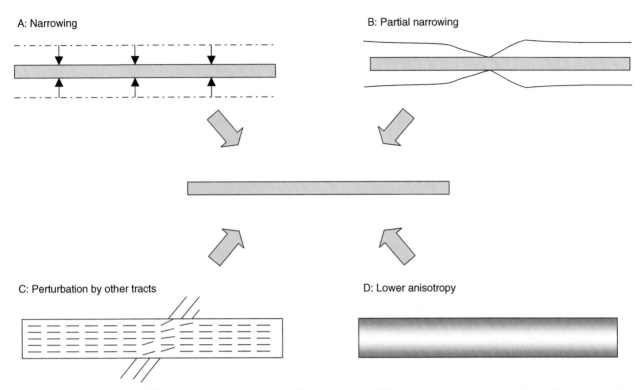

Fig. 9.26 Various possibilities that lead to a smaller tract size. When tractography results indicate that the size of a tract of interest has become smaller, this could indicate narrowing of the tract in its entire length as the tractography result indicates (A). However, the narrower reconstruction result could be a result of partial narrowing of the tract (B), perturbation by crossing fibers (C), or decrease of anisotropy below threshold level (D).

SNR of the data. If differences are found in tractography-based quantitative studies, it is very important to investigate the reason by inspecting the raw images such as anisotropy, orientation, and/ or tensor maps. Tractography is a screening tool that is sensitive to various anatomic events.

Chapter 10

Quantification approaches

In this chapter, quantification approaches are classified into two categories: first, DTI is used to improve existing quantification approaches; second, DTI can provide new anatomic information. These approaches are discussed in the following two sections.

10.1 IMPROVEMENT OF CONVENTIONAL QUANTIFICATION APPROACHES

10.1.1 DTI can improve manual ROI drawing

For quantitative studies of MRI, one of the most commonly used approaches is to manually define regions of interest (ROI). ROI could be placed on abnormal regions (e.g., T_2 hyperintensity areas) or predetermined anatomic regions with fixed ROI sizes. When ROIs are placed in the white matter, the criteria of the location and the size of the ROI are, however, not always clear. Because the white matter looks homogeneous in many conventional MRIs, it is difficult to define anatomic reasoning for the location and the size of the ROI. This could lead to poor reproducibility of ROI placement.

As shown in Fig. 10.1, DTI can provide an anatomic template for ROI placement. If the location of an ROI is arbitrarily defined in conventional MRI, it may contain multiple white matter tracts with different functions (Fig. 10.1B). Even a slight shift of location could lead to quantification of white matter regions with different anatomic meanings. If we have DTI data, we can design ROI placement (size and location) to define specific white matter tracts. This allows tract-specific quantification of MR parameters. We can also design the quantification in a hypothesis-driven manner; for example, if we suspect that the thalamo-cortical connections are affected by a disease, we can place an ROI on the corona radiata and control ROIs on unrelated tracts. Thus, the sensitivity and specificity of the ROI-based quantification could be improved, and the reproducibility of the ROI placement also increases.

This approach can be used with any MR image as long as they are coregistered to DTI, i.e., we can measure T_1, T_2, MTR, ADC, FA, etc., of specific white matter tracts that can be discretely identified in color-coded orientation maps. This is indeed a powerful technique enabled by DTI. However, there is a technical

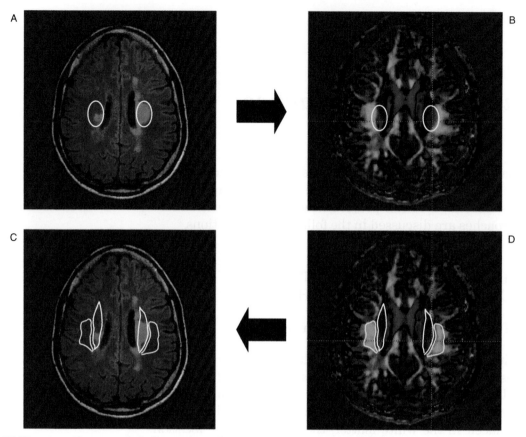

Fig. 10.1 Utilization of anatomic information of white matter for manual ROI drawing. In conventional MRI, placement of ROIs in the white matter is often based on arbitrary criteria (A), which may contain multiple white matter tracts with different functions (B). In this example, the oval-shaped ROI contains both the corona radiata (blue fiber) and the superior longitudinal fasciculus (green fiber). With DTI information, ROIs could be placed on specific white matter (C and D). The white ROIs define the corona radiata, and the yellow ones define the superior longitudinal fasciculus. In this example, the FLAIR image of a multiple sclerosis patient is used as an example of conventional MRI (A and C).

difficulty that we should be aware of. Because most DTI is based on EPI-based data acquisition, the images are distorted. We need to either undistort the DTI images or acquire all coregistered MRI with the same EPI-based method (all images are distorted in the same way). Due to the recent advent in MR technology, the amount of distortion in DTI has been reduced drastically, but misregistration could still be a significant issue. Before quantification is performed, it is essential to confirm that all MR images are aligned well.

Although the manual ROI-based quantification is a valid approach, it is not free from problems. Most notable is reproducibility when defining the boundary of a tract. In general, tubular-shaped tracts are easier to define than irregular-shaped ones. Selecting a slice orientation that is perpendicular to the tract (thus revealing its cross section) is a good approach. If the slice orientation is oblique or parallel to the tract of interest, slight differences in the tract angle could lead to a large difference in the tract size. In addition, such slice selection often leads to curvature of the tract within the slice

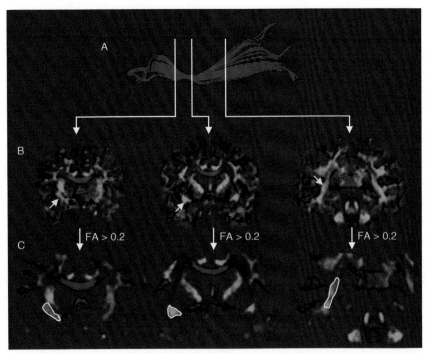

Fig. 10.2 A suggested scheme for manual ROI-based quantification. If the trajectory of a tract of interest is known (A), slices that are perpendicular to the trajectory are extracted from the core regions of the tract, where the tract is compact and can be discretely identified (B). The boundary is defined from the color and intensity. To assist in reproducible boundary definition, an intensity threshold can be used (C).

plane or divergence of the tract into cortical regions, which makes manual definition of the tract boundary difficult. If the trajectory of the tract is known, slices that are perpendicular to the trajectory can be extracted, as shown in Fig. 10.2.

When an ROI is manually drawn, the boundary is defined by the color (fiber orientation) and intensity (FA value) of the tract. The FA thresholding can assist reproducible definition. (Fig. 10.2C), although a tract of interest is often continuous to other tracts, and a simple FA thresholding may not clearly define a boundary. By combining both anisotropy and fiber orientations, the power of tract identification increases substantially.

10.1.2 Interpretation of ROI-based quantification needs caution

When a manual ROI approach is used to define a structure of interest, the contrast between the structure of interest and the surroudings is an important factor. When the structure has low contrast, ROI definition and the resultant intensity quantification become very sensitive to how the boundary is defined (Fig. 10.3). This problem is ameliorated if one uses an intensity threshold. However, the choice of an appropriate threshold value is another issue.

For example, if one is interested in quantification of FA and uses an FA value as a threshold, the quantification results depend on

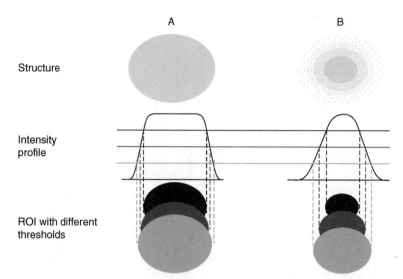

Fig. 10.3 Relationship between contrasts and ROI definition. When there is high contrast (A), differences in boundary definition have little effect on the ROI size. The ROI size becomes more variable when the contrast is low (B).

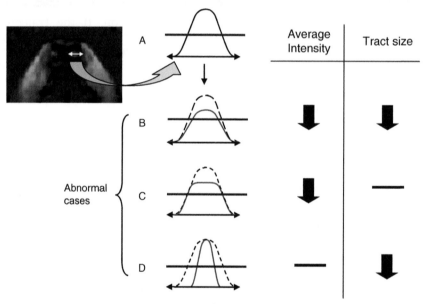

Fig. 10.4 Several possibilities of abnormalities in tract anisotropy. Hypothetical intensity (e.g., FA) profiles of the corticospinal tract along the double-headed arrow are shown in A–D. In (A), the black curve indicates a profile of FA across a tract of interest, and the blue line is a level of FA threshold to define the boundary of the tract. In case (B), the tract anisotropy decreases while the shape of the profile remains the same. If the same FA threshold is applied to define this tract, both the tract anisotropy and size would decrease. In case (C), anisotropy of the core of the tract decreases while that of the marginal area remains the same. In this case, average anisotropy of the tract decreases while the size of the tract remains unchanged. In case (D), the average intensity does not change significantly, while the tract size decreases.

the threshold of choice. This situation is the same (or worse) for the manual ROI method. We can obtain high FA by including only the core fibers or obtain low FA by including marginal regions. Apparently, consistency of the threshold is important.

Regardless of the ROI-drawing strategy, interpretation of the results needs caution. In Fig. 10.4, some hypothetical situations

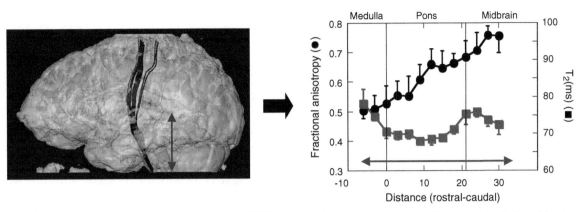

Fig. 10.5 Tractography-based quantification of MR parameters. Once a tract of interest is defined by tractography, its coordinates are superimposed on coregistered MR images. In this example, the corticospinal tract is reconstructed, and the path is superimposed on FA and T_2 maps. For each axial slice, average FA and T_2 values of pixels occupied by the tract are determined.

of tract abnormality and quantification results are shown. When the tract boundary is judged from intensity (either visually or using a threshold), both average intensity and tract size may or may not sensitively detect the change. These cases exemplify the importance of understanding the properties of our quantification tools.

10.1.3 Tractography can be used as a semiautomated region-growing tool

It is possible to identify a tract of interest in a slice-by-slice manner to define a 3D ROI along its trajectory, based on manual drawing, as described earlier. This can be more efficiently achieved by using tractography. For example, in Fig. 10.5, the corticospinal tract is defined by 3D tracking, and the coordinates are superimposed on FA and T_2 maps. Thus, tractography is used as a 3D region-growing tool to identify pixels that belong to a specific tract family.

Compared to the slice-by-slice 2D analysis, this approach can be more time efficient. In addition, unlike manual ROI, consistent criteria for the ROI boundary can be applied to the entire length of the tract. Furthermore, some white matter tracts can be defined only by tractography. For example, tractography can define the location of the corticospinal tract in the internal capsule and the corona radiata, which is otherwise difficult to define.

There are several disadvantages to this approach. First, tractography can be applied only to a subset of white matter tracts that can be reproducibly reconstructed. Although the reconstruction is automated, the results depend on the location and sizes of the initial ROIs and threshold values. Interpretation of the results needs caution because, as explained in Fig. 9.26, tractography results are influenced by many anatomical factors along its path. If a long section of a tract is reconstructed, the tract size could vary considerably among subjects. It is thus recommended that this approach be used for a short segment of well-defined tracts.

10.2   QUANTIFICATION OF ANISOTROPY AND TRACT SIZES BY DTI

10.2.1   Quantification of contrast and morphology: photometric or morphometric

When we quantify images, it is important to know whether we are interested in pixel intensity (contrast) or anatomy (morphology); the former is sometimes called photometry and the latter, morphometry (see Section 7.2). Quantification of FA and ADC is photometry. One of the most pressing issues of photometry is how to measure corresponding regions in different subjects. For morphometry, we are interested in the size and shape of specific anatomical structures, in which we often need to define a cluster of pixels that belong to the same structures. In this case, one of the most challenging questions is how to define the boundary of the cluster. Usually, we detect boundaries based on sharp contrast change. DTI can provide two new contrast mechanisms for boundary definition: diffusion anisotropy and fiber orientation. The latter information is unique to DTI and provides a sharp contrast at tissue boundaries. The downside of the orientation information is that it requires us to handle vector or tensor fields, which is often not straightforward.

10.2.2   Photometry: There are various types of quantification approach for contrasts

Anisotropy is a new DTI-based contrast. In Chapter 7, the biological meaning and interpretation of diffusion anisotropy are discussed. Because anisotropy maps are scalar images, anisotropy can be quantified by using conventional approaches. In Section 10.1, manual ROI-based and tractography-based approaches are described. A common issue of ROI-based approaches is the inverse relationship between reproducibility and localized information (Fig. 10.6). In an extreme case, we can identify the entire brain as an ROI. In this case, we can reproducibly define the ROI within a subject or across subjects, but there is no localized information. Stereotaxic approaches use several key anatomical landmarks such as width, length, and height of the brain and the anterior and posterior commissures to systematically divide the brain into smaller regions. This is a highly reproducible method and has superior localized information compared to the whole-brain ROI. We can further improve localization by manually defining specific anatomical structures. This approach has excellent localized information, albeit reproducibility is compromised. The smallest unit of localization is each pixel. Therefore, the other end of the extreme is pixel-to-pixel matching across subjects. Of course, this

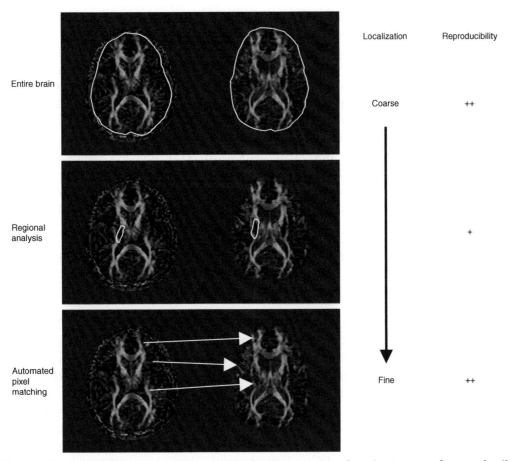

Fig. 10.6 Comparison of different contrast quantification approaches in terms of reproducibility and localization.

cannot be done manually (there are too many pixels in the brain). Voxel-based analyses perform the whole-brain pixel matching automatically using computer software developed for this purpose. Matching all pixels to corresponding pixels between two brains means transforming the shape of one to the other. Therefore, this procedure is often called "normalization," "transformation," or "brain warping." The reproducibility is high, and it has the highest possible localized information.

10.2.3 Morphometry: DTI reveals morphology of intra-white-matter structures, but quantification is challenging

By inspecting a color-coded orientation map, it is clear that it carries new information about intra-white-matter anatomy. In T_1- and T_2-weighted images, white matter appears to be a homogeneous structure, and all we could do is measure the entire volume of the white matter. With DTI, we can visualize the size and shape of individual white matter tracts, and this has great potential in many areas of brain research. For example, we know that brain volume changes during development and aging or under many pathological conditions. It is important to know if the entire white matter tracts change their volume proportionally or if there are specific tracts that account for the change (see, e.g., Fig. 11.6). If the white matter volume changes by 5%, this could be due to 50% volume change of a specific family of tracts. DTI can provide more specific and sensitive information about the morphology of the white matter. However, quantification of morphology is a challenging task in general. Even for T_1-weighted images, which typically have less than 10 MB of information, morphology analyses are not straightforward. DTI has six times more information than conventional scalar images. In the following sections, several quantitative approaches for white matter morphology analyses will be introduced with their advantages and disadvantages.

10.2.4 Morphometry: Manual ROI-based methods can be used to measure the size of specific white matter tracts

Similar to contrast quantification (Section 10.2.2), we can use a manual ROI-based approach for anatomical analyses. However, care must be taken in this approach; unless the observation plane is precisely perpendicular to the tract of interest, the cross-sectional area becomes sensitive to the angle of the tract with reference to the plane (Fig. 10.7). In this example, the superior cerebellar peduncle (SCP) is oblique to the plane. It is not uncommon for the angle of this tract with respect to the anterior–posterior commissure (AC–PC)–aligned axial plane to change when there is severe cerebellar atrophy. Even if the tract size does not change, alterations in the angle could lead to a different cross-sectional area. There are two ways of ameliorating this problem. First, the observation plane can be tilted perpendicular to the tract. Second, 3D ROIs can be defined from multiple observation slices. Similar to ROI-based contrast quantification, thresholding and slice-by-slice region-growing tools could be used.

10.2.5 Morphometry: Tractography can be used as a semiautomated tract definition

The 3D ROIs can be defined using a semiautomated tract definition tool such as tractography. Compared with slice-by-slice manual definition, the boundary of the tract can be defined in a more systematic and time-efficient manner. However, because tractography integrates all events that could influence the tracking along its trajectory (see Chapter 9, Fig. 9.26), we have to be cautious about the interpretation, i.e., the entire tract could look narrower if there is one narrow region or low-FA region along its tract. The two-ROI approach introduced in Chapter 8 is recommended, and the tract volume should be measured between the two ROIs for reproducible measurement. The closer the two ROIs, the lesser would be the influence of noise and other factors on the tracking results. However, the number of slices between the two ROIs

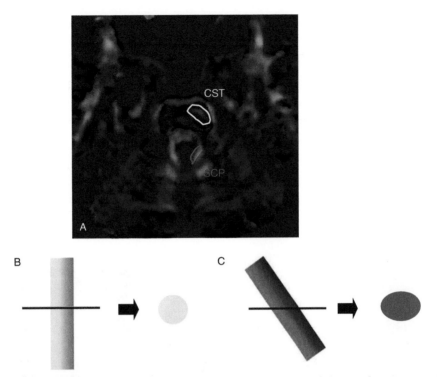

Fig. 10.7 Examples of ROI-based tract size quantification and its sensitivity to the tract angle. Image in (A) shows a color-coded orientation map of the human pons, in which the corticospinal tract (CST) and the superior cerebellar peduncle (SCP) are clearly identified. The CST is roughly perpendicular to the observation plane, which reveals the cross-section as shown in (B), while the SCP is oblique to the plane, which leads to the larger cross-section shown in (C).

is directly proportional to the tract volume. For example, if there are five axial slices between two ROIs in one subject and four slices in another, we expect a 20% difference in the tract volume. We can fix the separation of the two ROIs, but the measured region could shift along the tract. The two ROIs could be defined by anatomical landmarks to ensure reproducible placement but separation of the two landmarks may vary depending on patients. In this case, the measured volume may reflect the distance between the two landmarks rather than the volume of the tract of interest.

These possible sources of variation do not necessarily mean that volume measurements based on DTI/tractography are impossible or unreliable. Morphological analysis is a challenging task in general, and some of the issues raised earlier are not specific to DTI-based analyses. It is important to understand the various factors affecting the measurement results and use the knowledge for proper interpretation. It is advisable to measure tract sizes using multiple methods, including the voxel-based approach, which will be described in the following section.

10.2.6   Voxel-based analysis is becoming a popular tool, but has limitations

The voxel-based method is a very powerful tool for both photometry and morphometry. The advantages and disadvantages of various

TABLE 10.1 Comparison of several quantification methods with respect to reproducibility, localization, and accuracy

	Entire brain	Stereotaxic	Regional analysis			Voxel-based
			Manual ROI	Threshold/region growing	Tractography	
Reproducibility	+++	+++	+	++	++	++
Localization	+	++	+++	+++	+++	+++
Accuracy	+++	++	+++	+++	See Chapter 9	+-+++
Other features			Time consuming		Only for limited tracts	Minimum manual input

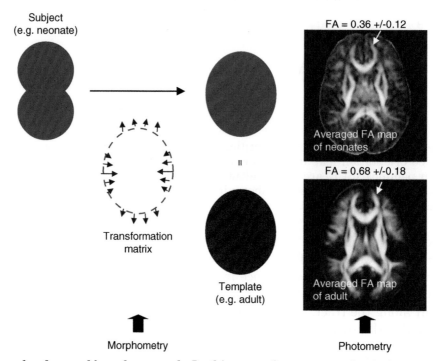

Fig. 10.8 An example of a voxel-based approach. In this example, a neonate brain is compared to an adult brain (template). After transformation, the shape of the neonate brain becomes the same as that of the adult brain. The transformation operation (transformation matrix) contains quantitative information about the shape difference (morphometry). The same procedure can be repeated for neonate and adult populations, and averaged maps for the two populations can be generated. In this example, average FA maps are created. Photometry (intensity comparison) can easily be performed by comparing pixel values at the same coordinates.

quantification approaches are given in Table 10.1 and Fig. 10.6. Among the three regional analysis tools, computer-aided methods could have higher reproducibility, as discussed previously.

Because of the high level of reproducibility, time-efficiency, and excellent localized information, voxel-based analysis is becoming a very popular tool. In Fig. 10.8, the idea of normalization is schematically shown. Suppose we want to compare brains of neonates and adults. Because their brain sizes and shapes are different, we first need to transform the neonate brains to an adult brain template. If we could make the shapes completely the same, quantitative information about the anatomical difference is stored in the transformation matrix. We can repeat this procedure for many neonate and adult brains. After this process, we can obtain many MR images (e.g., FA maps) with the same shape. We can then add all transformed images and calculate the average and standard deviation of the intensity of each pixel in the images. For example, in this figure, we created averaged FA maps of the neonates and adults. These averaged maps from two different groups can be simply superimposed for a pixel-by-pixel statistical analysis. It is very important not to be confused about the morphometric and photometric analyses at this point. If one is interested in brain anatomical difference, all the information is stored in the transformation matrix. So we need to analyze the transformation matrix. After the transformation, all brains look same, and they are of no use for morphometric analyses anymore. If one is interested in photometric analyses, transformed images are of interest because we no longer have to worry about where and how to draw ROIs to quantify pixel intensities. After normalization to the same template, we can ask, for example: what is the average FA of adults and neonates at the standard coordinates [94, 83]?

Of course, this works perfectly only when the image matching is perfect. If not, the transformation matrix cannot capture anatomical differences completely, and residual shape differences leak into the transformed images. Then, the subsequent pixel-by-pixel photometric analyses are not accurate anymore. In reality, transformation is not simple. For example, how can we find the corresponding pixels? Unfortunately, we have multiple solutions for finding the corresponding pixels, and the answer to the question could be different depending on which computer algorithm is used. To make matters worse, the answers that computers find may be wrong.

In Fig. 10.9, a simple diagram is used to explain the different levels of transformation approaches. Suppose we have two objects with different shapes; the simplest approach is rigid alignment, in which the object does not change its shape, and the difference is minimized by translocation (three modes: translocation along the x, y, and z axes) and rotation (three modes: rotation about the x, y, and z axes). The next level of transformation is linear transformation; for example, linear scaling (three modes) and shearing (three modes) are used to minimize the shape difference. To make the shape of the object even more similar to the target, we need to resort to high-dimensional nonlinear transformation.

The actual results of linear and nonlinear transformation and the average of five subjects are shown in Fig. 10.10. The averaged result of the linear transformation (Fig. 10.10B) looks much blurrier because their shapes are not transformed to completely the same shape. The sharper definition of the white matter by nonlinear transformation is a testimony to better registration.

By using high-dimensional transformation, it is possible to make the shape of two objects almost identical. This sounds good, but, apart from the long computational time, high-dimensional

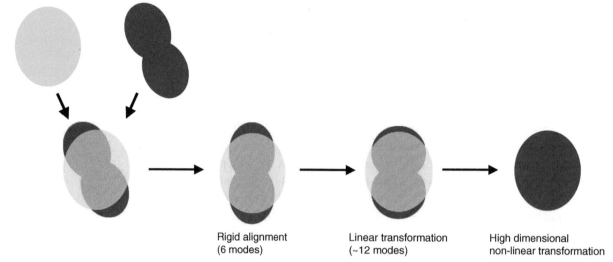

Fig. 10.9 A schematic diagram of low-dimensional to high-dimensional transformation approaches. The blue object is the target template, and the red object is transformed. In this example, there is no contrast within these objects. Therefore, the matching is performed entirely based on the edge information. For registration of real brains, matching of inside structures also needs to be maximized (see Fig. 10.13).

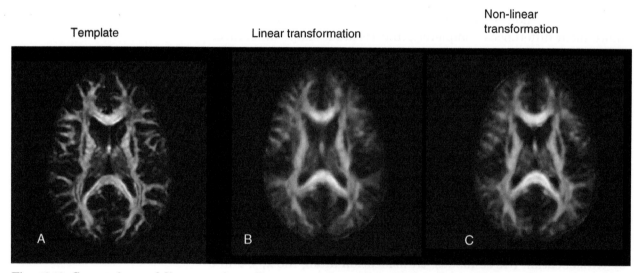

Fig. 10.10 Comparison of linear and nonlinear transformation. The image shown in (A) is the target template into which FA maps of five subjects are registered by linear transformation (B) or nonlinear transformation (C).

transformation has its own problems: the higher the dimensions, the more solutions we have. This point is illustrated in Fig. 10.11. There are many ways of matching the shapes of two different objects, and we do not know which one is anatomically correct. There may not even be corresponding structures (such as small gyri) between the two objects. In order to solve this difficult problem, high-dimensional transformation usually needs assumptions, such as minimum distances or minimum distortion. In addition, we do not know what we should do in the case of regions that have no features to match. For example, the outer shapes of the two objects in Fig. 10.11 have features to match, but what about the pixels inside

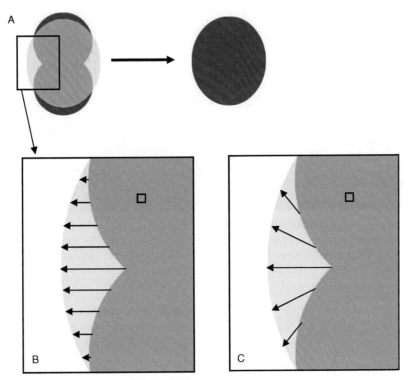

Fig. 10.11 There are multiple ways (B and C) to match the shapes of two objects (A). In this schematic diagram, a red object is transformed into a blue object.

the objects (e.g., the green pixel shown in Figs. 10.11B and 10.11C)? When we use linear transformation, the pixels inside an object move as a stretched rubber band would do; the transformation is global. In high-dimensional transformation, there are multiple options to move internal objects. If the boundary of the object is stretched or compressed, such movement could be confined to the local regions or can be extended globally using an interpolation scheme. We can come up with various solutions to this problem, but again we do not know which one is anatomically correct.

After discussing the issues related to the voxel-based approach, one very important question remains: is the method accurate? Here, accuracy means whether or not the computer software can find the corresponding anatomical structures. For example, we can easily define the anterior commissure at the mid-sagittal level manually. We have no doubt that all researchers can unequivocally identify its locations with high accuracy and reproducibility. Figure 10.12 shows the result of linear transformation of four normal subjects to a common template, and the resultant locations of the anterior commissures from these four subjects. If the transformation correctly finds corresponding pixels, the anterior commissures of all subjects should overlap that of the template. This result indicates that the accuracy of voxel-based analyses may not be as high as a manual-based ROI approach (Table 10.1).

One distinct advantage of the voxel-based approach is that it excels in whole-brain screening. The regional analysis is usually hypothesis driven, and we need *a priori* assumptions or knowledge

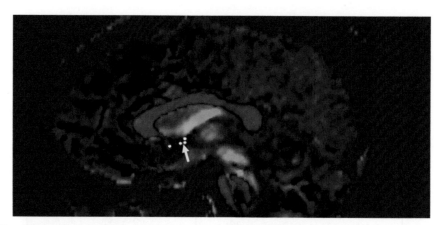

Fig. 10.12 Quality of image registration by linear transformation. Brains of four normal subjects are transformed to a template. The color map is that of the template subject, and the location of the anterior commissure is indicated by a yellow arrow. Resultant locations of anterior commissures of the four subjects are indicated by white dots.

to determine which regions to measure. The whole-brain analyses (including histogram analyses), stereotaxic analyses, and voxel-based analyses are suitable for examining the status of the entire brain and bringing our attention to sensitive brain regions. In conclusion, it is important to understand the characteristics of each approach and combine them effectively to answer specific biological questions.

10.2.7 There are several levels of brain transformation

The voxel-based analysis has been widely used for scalar images such as T_1-weighted images, and there are many types of software available for this purpose. We can use the same software for DTI, but there are several DTI-specific issues that we need to address. In Fig. 10.13, the matching of two scalar images is called Level 1, from which we can obtain a transformation matrix. Conventional T_1, T_2 images or FA maps can be normalized in this way. In these images, most of the anatomical features (contrasts) concentrate at the brain surface, ventricle shape, and gray matter–white matter boundary. Usually, there is not much contrast within the white matter. The transformation of the white matter could be inaccurate in this type of transformation. This has not been an issue for, e.g., fMRI, in which the white matter is not a structure of interest. If there are no visible structures or functional interest within the white matter, it is of no significance how the white matter is transformed. The situation is quite different for DTI, in which various white matter structures are visible, based on orientation (tensor) information. In Level 2 transformation, we can find where these intra-white-matter structures are moved after the transformation. At this point, there are two important issues. First, it is not straightforward to apply transformation to an orientation (tensor) field; this issue will be revisited in the following section. Second, suppose we could successfully transform the tensor field,

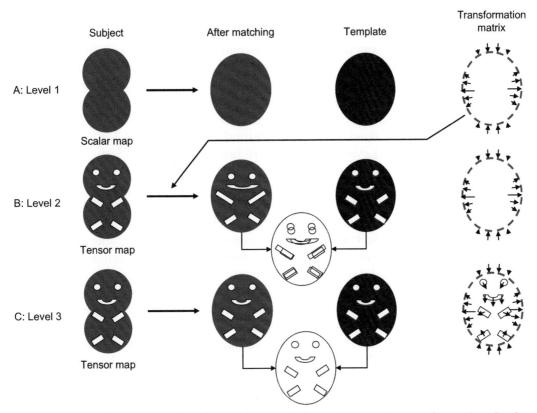

Fig. 10.13 Three levels of image matching for brain morphology. (A) Level 1 transformation: In the simplest case, we have two scalar images (e.g., T_1-weighted images, FA maps) to match. We can use conventional image matching software to obtain a transformation matrix. (B) Level 2 transformation: In a tensor field, we can identify various structures within the white matter, similar to the faces in the objects in this diagram. We can apply the same transformation obtained in the Level 1 transformation. This does not guarantee that the face elements align perfectly. (C) Level 3 transformation: Each feature in two tensor fields such as fiber orientations are aligned; this leads to the perfect matching of all anatomical features detected by DTI. The transformation matrix contains detailed morphological information of differences in intra-white-matter structures.

the resultant locations of white matter structures may not align to those of the template (Fig. 10.13B, Level 2). There are two ways of interpreting this mismatch. First, we can consider that it is a limitation (or error) of image matching. Second, we can treat it as anatomical differences between the two subjects after global anatomical differences (i.e., length, height, width, etc.) are removed.

It is a valid approach to use the Level 2 method and register multiple subjects to a template. Images in Fig. 10.10 are created using this approach. As expected, the matching of structures may not be perfect (Fig. 10.12), which leads to blurring after averaging a population. If there are two groups of populations, we can compare results between two averaged maps. However, we need to know several limitations in this approach. First, the transformation matrix does not contain morphological differences of intra-white-matter structures because it comes from the Level 1 method. Second, the amount of the resultant blurring (registration quality) depends on the registration software. Therefore, we cannot consider that the amount of blurring is purely due to anatomical differences between two subjects (or within groups). Third, if there are differences between two groups in the resultant averaged maps, say the cingulum is smaller (morphological difference) or its FA is lower (contrast difference), these could be due to a large anatomical variation in surrounding structures such as the ventricle and consequent poor registration of the cingulum, rather than differences of the cingulum itself. One interesting issue is that averaged FA maps obtained from the Level 1 approach and from the Level 2 approach are not the same. This issue will be discussed in more detail later.

In the more advanced approach (Level 3, Fig. 10.13C), anatomical features in tensor fields can be matched directly between two subjects. In this approach, registration software has to identify features such as a cluster of pixels with a specific fiber orientation and match these features between two images. This is similar to detecting face elements in the schematic diagram in Fig. 10.13C and matching eyes to eyes, nose to nose, and so on. If this works perfectly, we can obtain a new transformation matrix that contains anatomical differences in the white matter. The transformed images should completely match the template, and contrast quantification should be more accurate. For example, if the cingulum of a subject is smaller than that of the template, the registration software has to enlarge it to make it similar to the template. This information is stored in the transformation matrix, which we can examine later. Once the location and size of the cingulum becomes identical to those of the template, we can compare contrast (such as FA, ADC, T_2, etc.) of the cingulum readily without drawing ROIs. Unfortunately, we still have a long way to go to achieve Level 3 registration. We know that even the matching of scalar maps is not straightforward, as explained in Fig. 10.11 and Fig. 10.12. Matching of tensor maps is even more difficult. However, there is great potential in this research field.

10.2.8 How are tensor fields transformed?

In the Level 2 and Level 3 approaches in Fig. 10.13, tensor fields need to be transformed, which is not an easy task. This task is demonstrated in Fig. 10.14. When we transform a scalar image, our task is to identify where each pixel needs to go while maintaining the same intensity of the pixel. When we deal with a tensor field, we need to change the orientation of the tensor based on how adjacent pixels are moved (Fig. 10.14E). This is not a straightforward task because we have to consider how surrounding pixels are moved with respect to the pixel of interest. In Fig. 10.15, several hypothetical cases are used to explain this issue. Figures 10.15A and 10.15B show when sheering is applied to two situations with different fiber orientations. When fibers run horizontally, their angle needs to be rotated (Fig. 10.15A) while vertical fibers do not need to be reoriented (Fig. 10.15B). For global rotation (Fig. 10.15C), fibers need to be rotated regardless of their orientation. Fiber reorientation is not simple when nonlinear transformation is applied, in

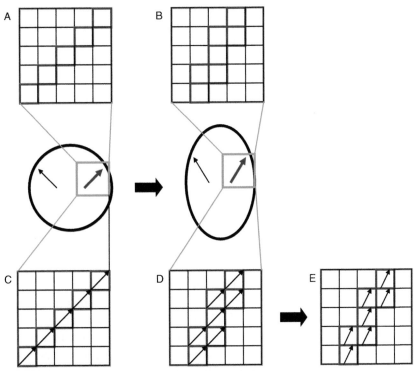

Fig. 10.14 Difference between scalar-based transformation (A and B) and tensor-based transformation (C and D). The circle object has a fiber running at 45° (A). After transformation, the object is stretched vertically, which changes the angle of the fiber steeper than 45° (B). In DTI, each pixel contains a tensor (C: a vector is used in this figure for simplification). If we simply move locations of pixels after stretching (D), the fiber angle information within each pixel remains at 45°, which does not reflect the new fiber angle. To solve this issue, we have to calculate the correct rotation angles from the amount of stretching (E).

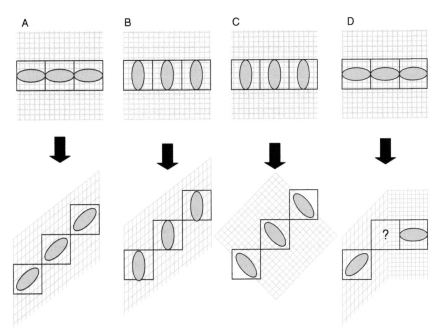

Fig. 10.15 Four hypothetical cases to explain the relationship between tensors and transformation. After transformation, the tensor in each pixel should be rotated in Case A and Case C, but not in Case B. It is not clear how Case D should be treated. It is also difficult to judge how the second and third eigenvectors should respond to transformation.

which the surrounding voxels may not move in a concerted manner (Fig. 10.15D). In addition, it is not clear what the orientation of the second and third eigenvectors should be. For example, in Fig. 10.15, we assume that the oval in each pixel represents orientations of the first and second eigenvectors. In Fig. 10.15A, after the sheering operation, we reoriented the first eigenvector (the longest axis). Because the shortest axis (the second eigenvector) is vertically oriented, should it remain vertical as in the first eigenvector in Fig. 10.15B? If the second eigenvector remains vertical, the angle between the first and second eigenvectors becomes no longer orthogonal, which is not allowed if we are using 3×3 tensor matrices. Therefore, the second eigenvector needs to be rotated with the first eigenvector, as shown in Fig. 10.15A. Thus, the second eigenvector is treated as a dependent of the first eigenvector. What if Fig. 10.15 shows the second and third eigenvectors of a 3D ellipsoid instead of the first and second eigenvectors? Should we reorient the second eigenvector responding to sheering and rotation operation? These are examples of issues we cannot easily solve (or to which there may not be a single correct answer). There are several mathematical models postulated to address these issues, and some are listed at the end of this chapter.

One frequently asked question is: can we perform all the transformations on raw diffusion-weighted images? Because raw images are based on scalar intensity, we do not have to deal with the difficult problems of tensor manipulation. However, there is also a complication in this approach, which is the b-table dictating applied gradient orientations. For example, the table consists of 30 gradient combinations if a 30-orientation scheme is used. Usually, the same gradient scheme is applied to all pixels for tensor calculation. However, if we deform the diffusion-weighted images, we need to deform the gradient orientations too, that is, we need to recalculate a new b-table for each pixel.

10.2.9 Population-averaged maps are created by adding transformed images

In Fig. 10.16, the flow of image registration is shown. The image transformation described earlier (also called image warping or normalization) is still the first step of the procedure, where a linear or nonlinear scheme is applied. After the transformation, we need to measure the degree of match with the template. Many methods have been proposed to assess the match of regular scalar images such as T_1, T_2, and FA maps. A function for measuring the degree of tensor-to-tensor match (Level 3) is a focus of active research. After many iterations, the process is complete and we can obtain a transformation matrix and a transformed image. If the matching is perfect, the transformed image looks exactly like the template, and all morphological information is stored in the transformation matrix. We can repeat this procedure for the all subjects in a group. We can add the transformed images of all subjects to create a population-averaged map.

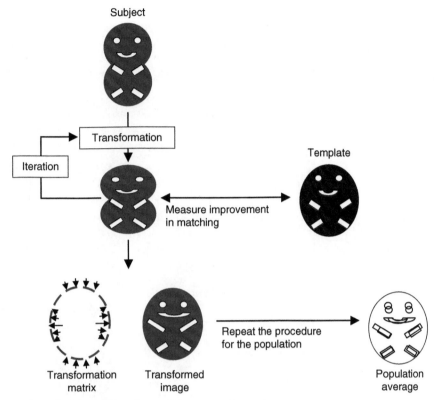

Fig. 10.16 A schematic diagram of the flow of image registration.

In the population-averaged map, there are two types of information. First, for each pixel, we can calculate average and standard deviation of the pixel values, such as T_1, T_2, MTR, ADC, and FA. These values can be directly compared between averaged maps from different groups at each standardized coordinate (see also Fig. 10.8). Second, it contains residual morphological differences that could not be removed during the transformation process. This leads to a "smearing" effect. The amount of the residual is larger in Level 1 and Level 2 in Fig. 10.13, in which white matter structures are not actively aligned among images. An example of the Level 2 method is shown in Fig. 10.17. As expected, the averaged map looks much less crisp than the template. In the Level 1 and 2 approaches, we can consider that the transformation is a procedure to remove differences in global brain sizes and shapes. Differences of white matter structures within a group are stored in the amount of smearing.

Once we get averaged maps from, e.g., patient and control groups, we can perform pixel-by-pixel analyses. For example, if the FA value of given standard coordinates of the averaged patient map is 0.4 ± 0.1 and that of the control map is 0.8 ± 01, we would suspect that there is significant loss of FA in the patient at that location. Similarly, if the fiber orientation of patient and control groups are $30° \pm 5°$ and $60° \pm 5°$ there could be reorganization of tract structures. However, as pointed out earlier, these differences in pixel values could also be due to differences in registration quality.

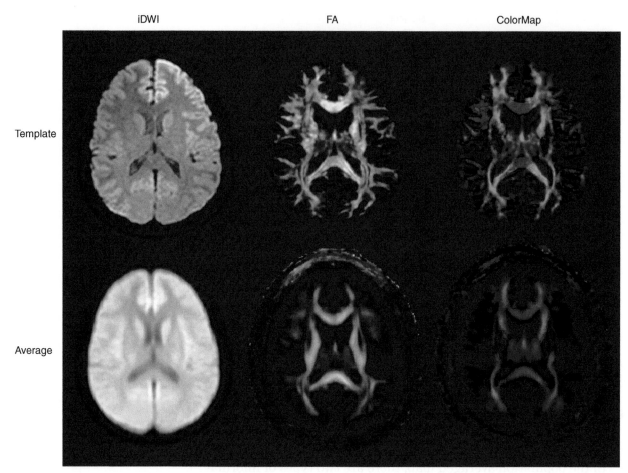

Fig. 10.17 Examples of population-averaged map using the Level 2 approach (Fig. 10.13). The population-averaged maps are created from data from ten normal subjects using a linear transformation.

For example, if the patient group has large inhomogeneity in the ventricle size, we would expect more smearing effect around the ventricle, which may alter, for example, FA values of the pixel of interest.

10.2.10 Tensor averaging is not simple; we may get different anisotropy values depending on how we average data

During the process of image registration, there are steps in which we have to average tensors. This is when we create a population-average map or when we interpolate two pixels during the transformation. In both cases, we have two or more pixels and calculate a mean tensor. Calculating a mean is an easy task for scalar maps. If there are two adjacent pixels of scalar values 10 and 20, the average of the two (or a pixel value interpolated right in the middle of the two pixels) would be 15. However, if we have two tensors, there is more than one way to calculate a mean. Fig. 10.18 illustrates this point.

When we have two adjacent pixels with different orientations, there are usually two possible anatomical configurations: first, the two pixels belong to one tract with a rather steep turn angle;

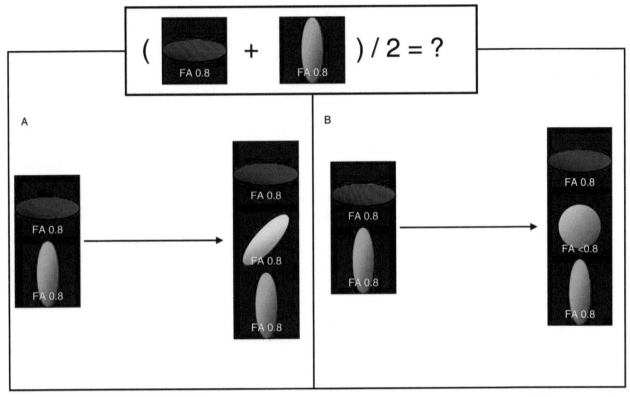

Fig. 10.18 The difficulty of averaging tensors. For example, to interpolate two tensors with different orientation, we can think of two ways, A and B. If we have *a priori* knowledge that there is one continuous fiber in these pixels, A is the better answer. If these two tensors represent unrelated fibers, Solution A creates a nonexisting fiber. Solution B would thus be a better answer. Namely, the middle pixel has partial volume effect.

second, these two pixels may belong to two unrelated tracts. In the DTI dataset, these two configuratoins degenerate and could result in the same tensor arrangement. Because we do not have *a priori* knowledge about the anatomical configuration, we have an ill-posed problem; i.e., we cannot unequivocally determine the underlying tract structures from DTI data. In this case, there are two ways of interpolating the two pixels. In the first solution (Fig. 10.18A), the anisotropy of the tensor is kept high, and it adopts a mean fiber angle, which creates three pixels with a smooth-angle transition. The second solution leads to a result similar to the partial volume effect (Fig. 10.18B); the interpolated pixel experiences reduction in anisotropy. The degree of the loss of anisotropy depends on the difference of the fiber angle. If the two pixels have exactly 90° difference, the first and second eigenvectors degenerate (the ellipsoid becomes a circle in the observation plane) and anisotropy is minimized. At a glance, the first approach may make more sense, but if the two pixels contain unrelated tracts, it would create a nonexisting tract between the two tracts. On the other hand, if the two pixels contain the same tract, the second approach leads to lower anisotropy every time interpolation occurs. In reality, transitions of angles between two pixels are rather small if they belong to the core regions of prominent white matter tracts (our imaging resolution is high with respect to their curvatures), and we can assume that loss of anisotropy is not significant. Most of the sharp-angle transition occurs between two unrelated fibers or peripheral white matter regions, in which the image resolution is not high enough to clearly resolve them, and partial volume effects most likely occur during data acquisition. Therefore, the second approach is a more conservative solution. Alternatively, we can perform interpolation in raw diffusion-weighted images that are scalar maps, and we can avoid the complication of tensor manipulation. This approach, however, leads to a result similar to the second approach because it essentially increases partial volume effects. In any case, we should realize that interpolation is not simple when we deal with a tensor field.

We face a similar issue when we calculate average tensor maps for group statistical analysis. Here again, we have two possible scenarios. First, we are certain that the transformation is of high quality and the pixel of interest belongs to the same tract in all subjects in the group. In this case, unlike image interpolation, we do not have to decide which method to adopt. We can calculate the results from either approach and perform statistical analyses. In the second case, we do have a certain degree of misregistration. In this case, we cannot assume that the same standard coordinates from subjects A and B correspond to the same tract. This is a very interesting issue, as shown in Fig. 10.19A. If we are averaging scalar maps such as FA (Fig. 10.19A), misregistration of, e.g., a pixel in the cingulum (FA = 0.8) to the corpus callosum (FA = 0.8) in a different subject leads to an averaged pixel with FA = 0.8. This is what routinely happens when we create scalar-averaged maps of the white matter: gross misregistration between unrelated white matter tracts cannot be detected. However, with the orientation information from DTI, we know more about white matter anatomy. With the tensor information (e.g., color-coded map), it is clear that anatomy is not correctly aligned (Fig. 10.19B). When we perform tensor averaging in this case, the first approach in Fig. 10.18A is clearly not appropriate, because we do not want to create a nonexisting fiber by merging the cingulum and the corpus callosum. The second approach, which is more conservative in the misregistered regions, would reduce anisotropy similar to a partial volume effect.

These considerations lead to an interesting fact: averaged FA maps created from scalar FA maps are essentially different from those created from averaged tensor maps. This is not to say that one is wrong and the other is right; there is no method that is absolutely correct. As long as the same quantification method is applied to all data, we can build a quantitative platform to compare groups. However, we can say that the latter approach (tensor averaging) is more sophisticated and has the potential to detect significant differences between two groups in anatomically correct locations. This is because misregistered regions have larger smearing effects (lose anisotropy) and become statistically inert (larger standard deviation in tensor properties).

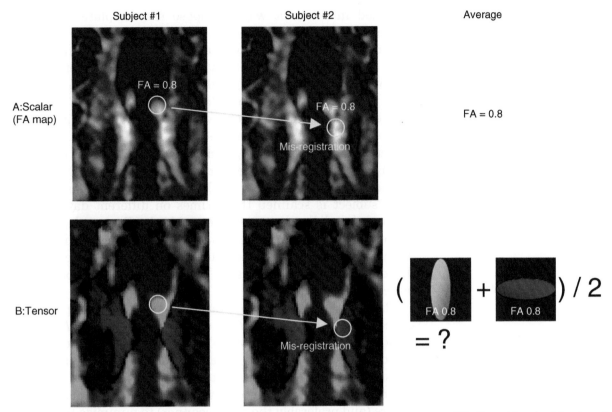

Fig. 10.19 Consequence of misregistration in scalar maps (A) and tensor maps (B). Even if a pixel in the cingulum is misregistered to the corpus callosum between two subjects, the averaged map could retain high anisotropy in the scalar maps. If the same misregistration happens in an averaged tensor map, we have several options to deal with the situation, as described in Fig. 10.18.

10.2.11 Image distortion in DTI needs special attention

As described in Chapter 6, EPI-based DTI suffers from image distortion. This distortion affects the normalization process at several different levels. First, we have to decide which images to use for normalization. If we use coregistered anatomical images such as T_1- and T_2-weighed images, the normalization process is not affected by the distortion but the transformation matrix is not completely applicable to DTI data. For example, we can perform Level 1 matching in Fig. 10.13 using T_1-weighted images and use the transformation matrix to normalize DTI data in Level 2 matching. In this case, distortion is not an issue in the Level 1 matching. However, the resultant transformation matrix cannot accurately normalize the distorted DTI. As an alternative approach, we can use DTI-based images such as FA maps or the least-diffusion-weighted images for normalization. Because all DTI-derived images from the same subject are distorted in the same manner, a transformation matrix obtained from one of the images (e.g., the least-diffusion-weighted image) is applicable to all the other images such as FA, trace, and tensor maps from the same subject. There is, however, an issue of cross-subject differences in the image distortion. Namely, the target and subject brains are distorted in different ways. Suppose the template and subject images have anatomy of A_T and A_S and distortion of D_T and D_S, respectively. The DTI image of the template is expressed as $[A_T + D_T]$. When the subject image is normalized, not only true anatomical differences (ΔA) but also the difference in the distortion term D (ΔD) is considered as a part of anatomical difference. In addition, because the template contains the D_T term, so do the resultant averaged maps.

The impacts of ΔD and D_T terms on our study depend on whether we are conducting photometry or morphometry studies. For morphometry studies, the term ΔD leads to inaccurate results. If the same subject is scanned three times on different days, it is not unusual that differences are observed in the D term, depending on the head positions and sinus conditions ($\Delta D \neq 0$). In this case, the normalization-based morphometry studies would detect ΔD from the three images of the same subject, which is apparently an inaccurate result.

The effect of ΔD terms may be not as significant for photometry studies, in which ΔA (and ΔD) is usually discarded. The D term, if it is too large, would have adverse effects on both morphometry and photometry studies because it may alter signal intensity and reduce registration quality. Because D term is highly sensitive to imaging parameters, it is also important to have the same imaging parameters (and preferably the same scanner) to minimize the ΔD. Due to recent advances in hardware and image reconstruction techniques, the amount of distortion (D) is steadfastly decreasing. There are also effective post-processing distortion correction methods. It is important to minimize the amount of distortion (D) and to carefully interpret its impact on normalization-based study results.

Chapter 11

Application studies

11.1 BACKGROUND OF APPLICATION STUDIES OF DTI

11.1.1 Application studies can be classified into qualitative and quantitative studies

Application studies can be divided into qualitative and quantitative studies to examine how DTI can contribute to or change the way we perform MRI-based research. In many cases, we can probably also call them radiological diagnoses (qualitative studies) and neurological research (quantitative studies). In the former studies, we need to diagnose the status of each patient. MR is required to have the power to differentiate abnormal and normal unequivocally for each individual basis. In this type of application, the degree of abnormality needs to be much larger than the amount of individual variation (Fig. 11.1A). In neurological studies, quantification is an essential part of the study. In many neurological studies, MRI is often not sensitive enough to differentiate normal from abnormal on a single-subject basis (Fig. 11.1B). Even in such a situation, MRI could be an important tool if it can detect differences in group average, because it could be used to infer disease mechanisms or to monitor effects of therapy, for example. Quantification is also essential if we want to perform correlation studies between an MR parameter and another parameter (MR or non-MR parameters), such as cognitive and motor functions. In the following sections, we will discuss how DTI can contribute to these different types of study.

11.1.2 Four types of new information provided by DTI

For application studies, it is important to know what kind of new parameters DTI can provide. We can probably classify new information into four categories: new contrasts, white matter morphology, refined information about anatomical locations, and connectivity (Fig. 11.2). Among new contrasts, FA is one of the most widely used parameters. For white matter morphology, DTI can delineate sizes and shapes of specific white matter tracts. It can also reveal deformation of tracts due to tumor growth. Needless to say, connectivity is new and unique anatomical information we can study using DTI. The "location" information may need some further explanation.

Conventional MRI can detect brain abnormalities through contrast changes in T_1 and T_2. If abnormality is located in the

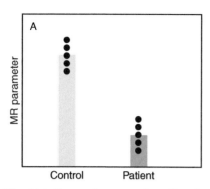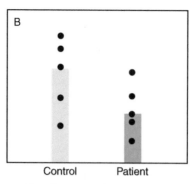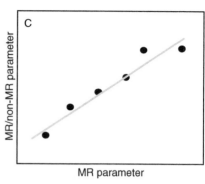

Fig. 11.1 Several types of application studies. In case (A), the degree of abnormality is much larger than the amount of individual variation. In case (B), the individual variation is larger than the degree of abnormality. In this case, data from a single person cannot be used to diagnose abnormality. In case (A) and (B), the vertical axis is an MR parameter, which could be image intensity (e.g., T_1, T_2, FA, ADC, etc.) or morphological parameters (e.g., the size of white matter tract). In case (C), an MR parameter is correlated to another parameter such as clinical data. In cases (B) and (C), quantification is essential.

white matter, we sometimes do not know its exact locations in terms of white matter anatomy and functions. For example, in the central semiovale, a 1 cm shift in the location of T_2 hyperintensity could involve totally different white matter tracts with different functions (Fig. 11.3). DTI can provide an anatomical template of the white matter, which can improve our ability to better localize contrast and/or anatomical alterations.

In this chapter, our main interest is to think how these types of new information can help us improve our ability to better delineate brain abnormalities. The improvement could be better segregation of patient and control groups (Figs. 11.4A, 11.4B, and 11.4C) or better correlation between MRI measurements and, e.g., functional outcomes (Figs. 11.4D and 11.4E).

11.2   EXAMPLES OF APPLICATION STUDIES

In the following sections, I would like to use some DTI results as examples to explain how the four types of new information (Fig. 11.2) can be used. For comprehensive reviews of current DTI application studies, readers are recommend to refer to an issue of *NMR Biomed.*, Volume 15, 2002, in which the entire volume is dedicated to reviews of DTI.

11.2.1   DTI can delineate gloss abnormality in the white matter anatomy better than conventional MRI

In Fig. 11.5, color-coded orientation maps of two cerebral palsy (CP) patient are compared with an age-matched control, together with coregistered T_2-weighted images. To diagnose these patients as CP, conventional MRI is sufficient. The enlarged ventricle immediately reveals that the patients have periventricular leucomalacia. Volumes of the white matter in the periventricular regions are severely reduced. What DTI can provide us with is a detailed anatomy of the

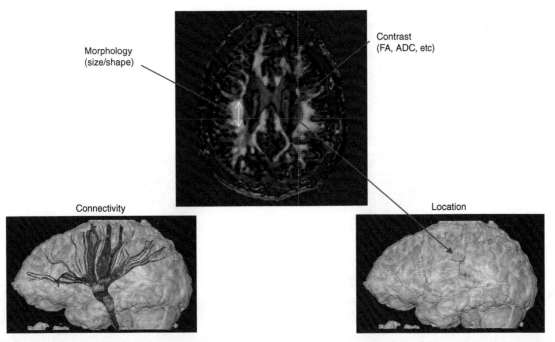

Fig. 11.2 Four types of new information that DTI can provide: new contrasts, white matter anatomy, refined information about locations, and connectivity.

affected white matter regions. In this periventricular region, there are four major white matter tracts: the corpus callosum (CC, red), the cingulum (CG, green), the corona radiata (CR, red), and the superior longitudinal fasciculus (SLF, green). If the volume of the white matter reduces, there are two possibilities. First, all constituent tracts reduce their volume proportionally. Second, there are specific tracts that are affected and account for most of the volume loss. In these patients, we can see that the latter is the case; the corpus callosum and the corona radiata are almost wiped out in the posterior region of the brains. This is a new type of anatomical information that cannot be obtained from conventional MRI. If one is interested in the prognosis of functional outcomes for babies with CP, imaging of specific axonal loss is likely to provide crucial information to improve our prognosis ability. This is an example of improvement described in Figs. 11.4D and 11.4E.

Another example of CP patients is shown in Fig. 11.6. In this figure, T_1-weighted images of two CP patients of the pons (Figs. 11.6A and 11.6C) do not show apparent abnormality. The pons of Patient 2 is smaller than that of Patient 1, but we cannot immediately tell if the small size of pons of Patient 2 is beyond the normal range. Namely, T_1-weighted images provide a type of information indicated in Fig. 11.4A. On the other hand, DTI-based color maps clearly show that the corticospinal tracts of the Patient 2 have severe atrophy, especially in the right side. This is an example in which DTI improves our current diagnosis ability from Fig. 11.4A to Fig. 11.4C. Because its ability to detect abnormality is so clear, the method can be used as a clinical qualitative diagnosis.

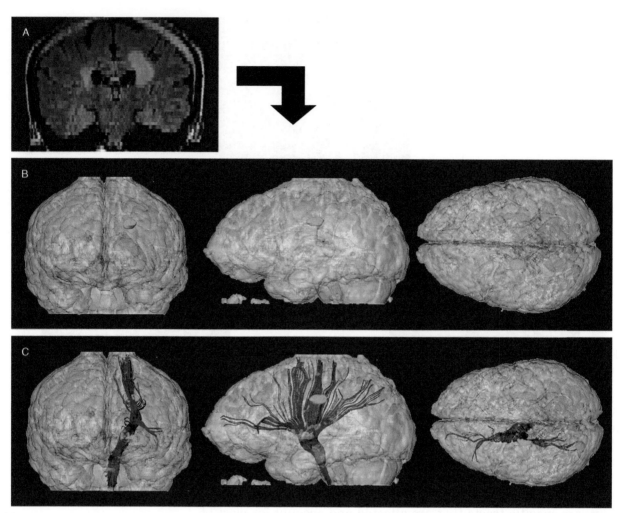

Fig. 11.3 An example of refined location information using DTI. Plaques in multiple sclerosis patients can be identified using a T_2 contrast (A). In this case, there is a large plaque in the central semiovale, which is defined three-dimensionally (green structure in B). If we have coregistered DTI data, we can locate the plaque on the corona radiata (purple structure in C) that contains the corticospinal tract (blue fibers in C).

11.2.2 Tractography can delineate deformation of white matter tracts in brain tumor patients

When there is severe deformation of white matter anatomy, it is often difficult to understand the anatomy based on slice-by-slice inspection of 2D images. Tractography can help in such a situation. Figure 11.7 shows an example in which tractography was used to decipher complicated alteration of white matter anatomy due to callosal agenesis. Without the help of tractography, it would have been much more difficult to understand the anatomy. Of course, the tractography result indicates only the macroscopic arrangement of white matter tracts and it by no means determines complete axonal connections in these patients. Still, it provides precious information to differentiate between various types of agenesis and their mechanisms.

Macroscopic alteration of white matter anatomy occurs commonly with brain tumor. DTI may not be a very effective tool

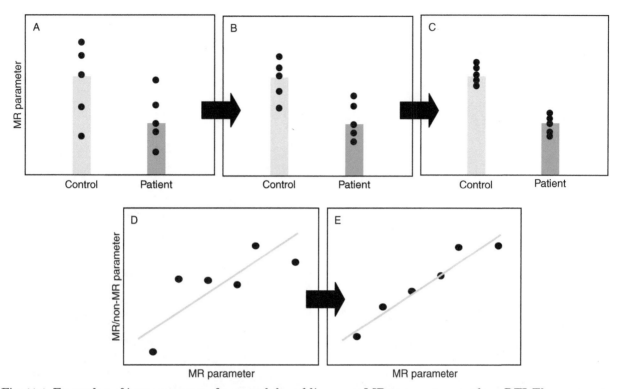

Fig. 11.4 Examples of improvement of research by adding new MR parameters such as DTI. The parameter helps better segregation of subject groups (A, B, and C) or better correlations with, for example, functional outcome.

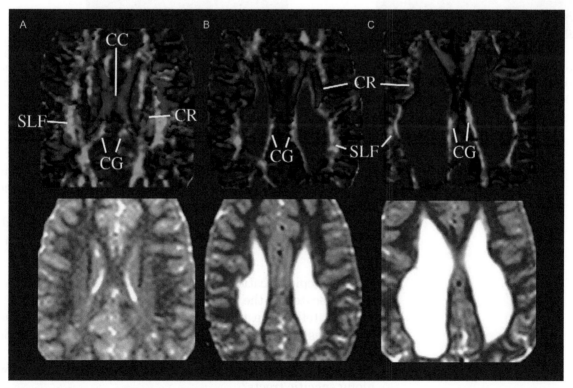

Fig. 11.5 Comparison of DTI-based orientation maps (upper row) and conventional T_2-weighted images (bottom row) of a 5-year-old control subject (A) and two cerebral palsy patients (B and C). CC: corpus callosum, CG: cingulum, CR: corona radiata, and SLF: superior longitudinal fasciculus. Images are reproduced from Hoon et al., *Neurology*, 59, 725, 2002, with permission.

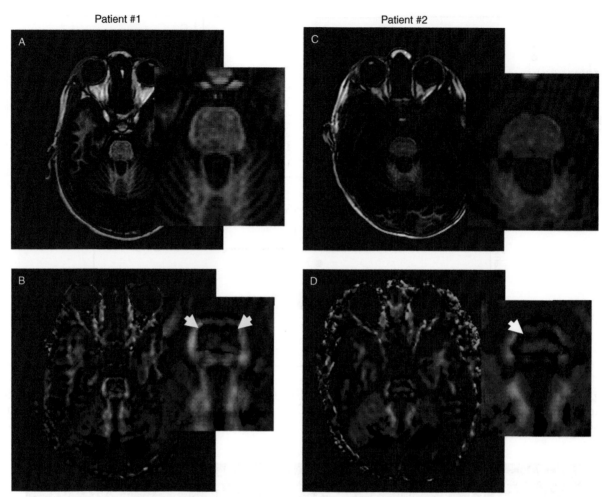

Fig. 11.6 Comparison of T_1-weighted (upper row) and color-coded maps (bottom row) of two CP patients. Pons regions are magnified. Arrows indicate locations of the corticospinal tracts. Images are provided courtesy of Alexander Hoon and Elaine Stashinko, Johns Hopkins University, Baltimore, Maryland.

to image the tumor itself, because a tumor in the white matter tends to lose FA, and it is often difficult to differentiate adjacent normal gray matter from tumor tissue. However, we expect that DTI can better delineate relationships between the tumor and the adjacent healthy white matter structures. There are three questions we can ask:

1. Can we depict locations of important white matter tracts that are dislocated by the tumor?
2. Can we tell if the tumor is growing by pushing the nearby structures or by invading without significant dislocation of nearby structures?
3. Can we tell the orientation of tumor growth if it is growing along specific tracts?

These three questions are discussed in greater detail in the following text.

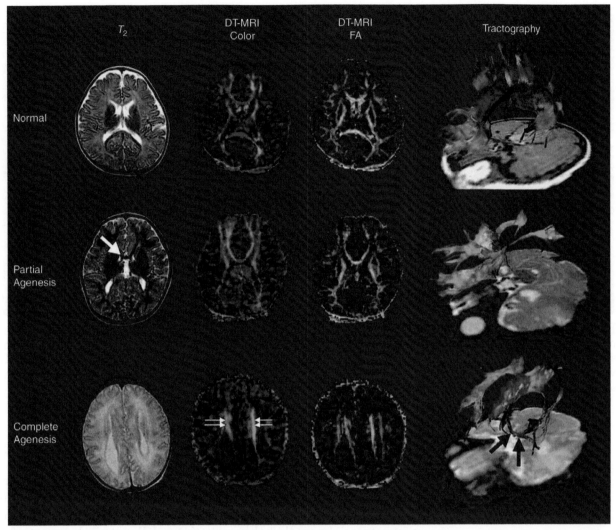

Fig. 11.7 Reconstruction of callosal connections in patients with callosal agenesis. Three-dimensional tract reconstruction is a powerful tool to decipher severely altered white matter structures. In the partial agenesis patient, a small amount of the corpus callosum is left, which connects the frontal lobe. Images reproduced from Lee et al., *AJNR*, 25, 25–28, 2004.

Imaging of dislocated white matter

For brain surgery, it is very important to avoid several key functional regions of the brain; these include motor, auditory, language, and vision fields. We have a fair amount of knowledge about functional maps of the cortex, but we do not have equivalent maps of the white matter. The locations and functions of the cortex can be deduced from the folding patterns of the cortex but the white matter appears as just a homogeneous structure during surgery. Even if we avoid injury to an important cortical area, the patient could lose function if the white matter tract responsible for the function is cut. Therefore, the identification of motor, language, auditory, and visual pathways is very important for brain surgery. This is especially so if a tumor mass dislocates these pathways and our standard knowledge of white matter anatomy of the normal

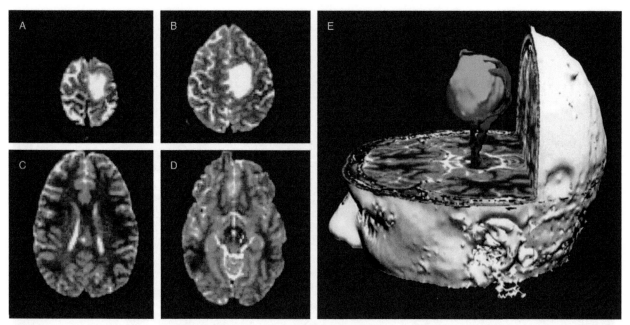

Fig. 11.8 Identification of motor pathways in a patient with astrocytoma. The shoulder (red) and wrist (blue) motor cortices were identified by intraoperative cortical stimulation, and motor pathways were reconstructed using tractography. The locations of pathways are shown at four axial levels in (A)–(D). The 3D view (E) shows the pathways in relation to the tumor (green). Images are reproduced from Berman et al., *J. Neurosurg.*, 101, 66, 2004 with permission.

brain cannot be applied. In Fig. 11.8, mapping of the motor pathway in a tumor patient is demonstrated.

This type of information could be very important for neurosurgeons to decide a surgical route, but there are two issues. First, we do not know how accurate the tracking results are; even if they are accurate, we know that they sometimes delineate only a part of a tract of interest. For example, tractography can usually depict the motor pathway only to the medial regions of the cortex, and it lacks trajectories to the lateral regions, probably because the motor pathways are outnumbered by massive association fibers. For surgery, it is vital that we be able to tell whether the motor pathway is posterior or anterior to, or right or left of, the tumor. It is likely that tractography does have this level of accuracy. However, we need a systematic study to correlate the tractography results and the surgical outcome to confirm that presurgical delineation of important white matter tracts does improve the outcome.

The second issue is our lack of knowledge about the relationship between white matter tracts and functions. There are many white matter tracts that have been described by anatomists and which have recently been noninvasively visualized by DTI. However, information about the functions of white matter tracts is surprisingly scarce. For example, if we could identify that the motor pathway is posterior to a tumor and decide to remove it from the anterior direction, we assume that the white matter anterior to the tumor in less important. This assumption may not hold true, depending on the location of tumor. It is important to assign functions to white matter tracts. DTI provides a foundation to substantially improve our understanding of anatomy–function correlation by providing a fine map of the white matter.

Differentiation of types of tumor and growth orientation

Malignant tumors tend to invade nearby structures without causing significant structural deformation. The relationship between the tumor growth and deformation of the surrounding structures can be appreciated in the color maps (Fig. 11.9).

Jellison et al., in their review paper (AJNR, 25, 356, 2004), classified the ways in which tumor interferes with axonal anatomy,

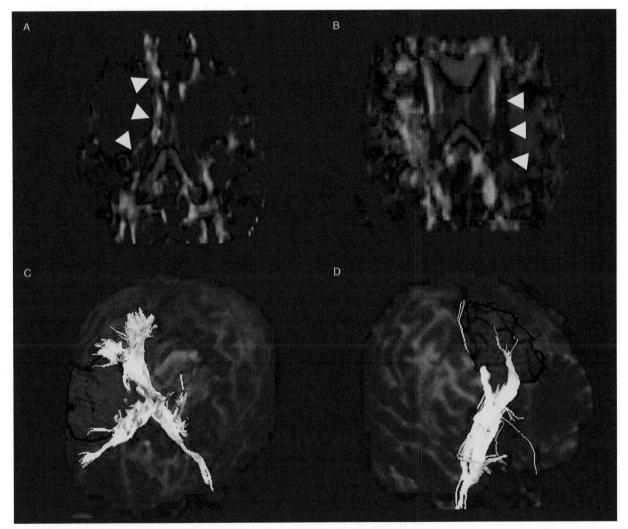

Fig. 11.9 Examples of tractography studies of two brain tumor patients. These are two cases of astrocytoma. In Case 1, the tumor grows discretely, pushing the surrounding white matter tracts (A) including the corona radiata (indicated by yellow arrows). In Case 2, there is no discernable deformation in the corona radiata (B). Three-dimensional reconstruction of the tumor (red) and the corona radiata (yellow) clearly delineate the relationships between the tumor and the tract. In Case 1, the tumor dislocates the corona radiata medially, and the tract envelops the tumor (C). In Case 2, the corona radiata projects into the tumor mass (D). The corona radiata contains such important tracts as the motor pathway. Images reproduced from *Annals of Neurology*, 51, 377, 2002.

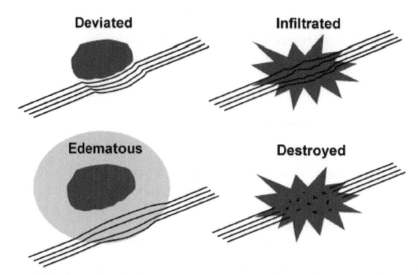

Fig. 11.10 Potential patterns of white matter fiber tract alteration by tumor. Images are reproduced from *AJNR*, 25, 356, 2004.

as shown in Fig. 11.10. Depending on information about white matter deformation (deviated cases), loss of diffusion anisotropy (infiltrated or destroyed cases), and T_2 hyperintensity (edematous), DTI could give us more information about tumor growth than conventional MRI alone.

Also, it is not uncommon to encounter a tumor which has one side that invades the white matter while the other side grows by pushing the surrounding structures with a clear boundary. There is the possibility that DTI can three-dimensionally map the degree of malignancy and predict the growth pattern and its orientation. There are two issues related to this capability of DTI. First, it still remains to be seen whether this type of information can actually improve diagnosis or therapeutic measures. For example, we may need to attack the tumor more aggressively (e.g., by surgery or radiation) in the area where the tumor is invading the surrounding structures. This type of hypothesis needs to be confirmed in future studies. Second, anatomical information similar to Fig. 11.9A and 11.9B can also be appreciated from conventional MRI to some degree. DTI provides more detailed information of nearby white matter anatomy. However, we need to study if this improved information makes any real difference in clinical practice. This again needs more investigation and should be rigorously tested against conventional MRI.

11.2.3 DTI can delineate anatomy of immature brains better than conventional MRI

Anatomic information of relaxation-based contrasts relies heavily on the amount of myelin. The human brain has only a small amount of myelin at birth and myelination takes place mostly in the first 2 years of life. As a result, T_1 and T_2 contrasts undergo a large amount of contrast change in this period (Fig. 11.11). As a

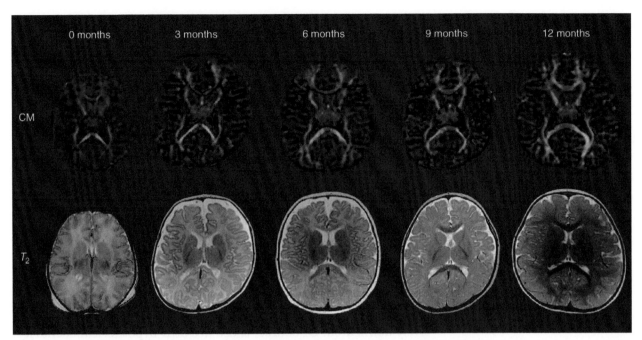

Fig. 11.11 Changes of imaging contrasts during the first 12 months after birth. The upper row shows DTI-based color maps (CM), and the bottom row shows T_2-weighted images. Images are provided courtesy of Laurent Hermoye, Saint-Luc University Hospital, Brussels, Belgium.

matter of fact, the gray matter–white matter contrast is inverted (bright white matter and dark gray matter) in T_2-weighted images at birth, which gradually approaches that of matured brains (dark white matter and bright gray matter) as the myelination progresses. During this process, there is inevitably a time period when even gray matter–white matter separation is not clear (e.g., the internal capsule at 3–6 months of age in Fig. 11.11). In Fig. 7.5 of Chapter 7, it is shown that anisotropy of white matter tracts of mouse embryonic brains is high even if the tracts are not yet myelinated. This point is also appreciable in the human data shown in Fig. 11.11.

The fiber orientation information (color information in Fig. 11.11) is relatively stable during postnatal development, suggesting that most axonal structures are already formed at birth. This ability of DTI to reveal immature white matter structures is important because DTI can provide an anatomical template to locate specific white matter tracts, which allows us to monitor maturation process (FA and T_2 measurement) in a tract-specific manner.

11.2.4 Status of specific white matter tracts can be monitored

In Chapter 7 and Chapter 10, the idea of tract-specific quantification is described. Application studies of this approach can be divided into two classes. In the first class, there are MRI-detectable abnormalities in the white matter, such as lesions with T_2 hyperintensity (e.g., adrenoleukodystrophy and multiple sclerosis). In this type of disease, we can use white matter anatomy information from DTI in one of two ways: (1) start from the identification of lesion locations based on conventional MRI and ask, "which white matter tracts are involved"; or (2) identify white matter tracts of interest using DTI data and ask, "is this tract affected by the lesion" by superimposing the tract coordinates on MR parameter maps. For the latter approach, we can, for example, establish protocols to determine the coordinates of motor and visual pathways and measure T_2, FA, and ADC along the pathway. Both are valid approaches and, of course, we can do both. We expect these approaches to increase anatomical specificity of our findings by adding DTI information (even though abnormality itself can be detected by conventional MRI for this class of disease). We can also expect the sensitivity to increase. For example, if a disease progresses along a tract, the core of the lesions could be detected by abnormally high T_2. However, peripheral regions may have elevated T_2 only by, for example, 5%, which is difficult to detect using slice-by-slice inspection of T_2 maps. If we can plot T_2 along the affected tract, it is much easier to identify the leading edge of the abnormality along the tract. Progression and effects of therapy could be more sensitively delineated. Of course, these are still hypothetical scenarios. This area of research has just begun, and we have yet to see if this kind of idea can be proved to be useful.

In the second class, patients are MRI negative. If this is the case, sensitive screening methods to survey the entire brain become useful. The use of DTI may help us in the following two respects. First, anisotropy could be a more sensitive parameter than other contrast mechanisms (this point is discussed in Chapter 7). Second, by allowing us to focus on the most eloquent pixels in the brain, DTI increases the sensitivity of abnormal detection. This is a hypothesis-driven approach. For example, if one hypothecates that limbic fibers are most affected in Alzheimer's disease patients, we can extract those pixels that belong to limbic fibers, such as the cingulum and fornix, and their T_2, ADC, and FA, for example, can be quantified. Again, this type of approach has begun to be used recently, and its usefulness is yet to be evaluated. However, it is a fact that DTI can provide a better white matter anatomical template, and it is important to try to harness this new information in order to improve our ability for sensitive and specific quantification.

11.2.5 Structure–function relationship studies will be an important future effort

For many studies introduced earlier, the structure–function relationship of the white matter is of essential importance. As already mentioned, we have a functional map of the cortex, in which the folding pattern is an important clue for structural localization. The fact that DTI can show many anatomical units within the white matter could provide us with the important "clue" for performing the structure–functional relationship of the white matter and create its functional map (i.e., location information in Fig. 11.2 and Fig. 11.3). If we have an elaborate structure–function correlation map, we can tell the neurosurgeon, "if this part of the white matter is cut, this patient will lose verbal recognition;" or a neurology doctor can tell stroke patients, "You still have the pyramidal tract left. Based on our previous cases, it is likely that you will regain your motor function." In reality, there may not be a clear structure–function relationship in many regions of the white matter. However, finding even a loose relationship is also an important part of the research.

A correlation study requires many state-of-the-art image analysis techniques, especially group analysis techniques. We can use stroke, lobectomy, or other cases with brain lesions, from which we can investigate which tracts are injured and which functions are lost. Each subject may have multiple injuries in the gray and white matter and, thus, the functional loss cannot be assigned to one specific tract. We need to create a database and perform statistical analysis to find the correlation between fibers and functions. DTI-aided brain functional mapping is an exciting future research field.

References and Suggested Readings

PREFACE

1   Basser, P.J. and D.K. Jones. 2002. Diffusion-tensor MRI: theory, experimental design and data analysis — a technical review. *NMR Biomed.* **15**: 456–467.

2   Basser, P.J., J. Mattiello and D. Le Bihan. 1994. MR diffusion tensor spectroscopy and imaging. *Biophys. J.* **66**: 259–267.

3   Basser, P.J., J. Mattiello and D. LeBihan. 1994. Estimation of the effective self-diffusion tensor from the NMR spin echo. *J. Magn. Reson. B* **103**: 247–254.

4   Beaulieu, C. and P.S. Allen. 1994. Determinants of anisotropic water diffusion in nerves. *Magn. Reson. Med.* **31**: 394–400.

5   Chenevert, T.L., J.A. Brunberg and J.G. Pipe. 1990. Anisotropic diffusion in human white matter: demonstration with MR technique in vivo. *Radiology* **177**: 401–405.

6   Doran, M., J.V. Hajnal, N. van Bruggen, M.D. King, I.R. Young, et al. 1990. Normal and abnormal white matter tracts shown by MR imaging using directional diffusion weighted sequences. *J. Comput. Assist. Tomogr.* **14**: 865–873.

7   Douek, P., R. Turner, J. Pekar, N. Patronas and D. Le Bihan. 1991. MR color mapping of myelin fiber orientation. *J. Comput. Assist. Tomogr.* **15**: 923–929.

8   Hsu, E.W. and S. Mori. 1995. Analytical interpretations of NMR diffusion measurements in an anisotropic medium and a simplified method for determining fiber orientation. *Magn. Reson. Med.* **34**: 194–200.

9   Le Bihan, D., E. Breton, D. Lallemand, P. Grenier, E. Cabanis, et al. 1986. MR imaging of intravoxel incoherent motions: application to diffusion and perfusion in neurologic disorders. *Radiology* **161**: 401–407.

10   Le Bihan, D., R. Turner and J. MacFall. 1989. Effects of intravoxel incoherent motions (IVIM) in steady-state free precession (SSFP) imaging: application to molecular diffusion imaging. *Magn. Reson. Med.* **10**(3): 324–337.

11   Moonen, C.T.W., J. Pekar, M.H. de Vleeschouwer, P. van Gelderen, P.C.M. van Zijl, et al. 1991. Restricted and anisotropic displacement of water in healthy cat brain and in stroke studied by NMR diffusion imaging. *Magn. Reson. Med.* **19**: 322–327.

12   Moseley, M.E., Y. Cohen, J. Kucharczyk, J. Mintorovitch, H.S. Asgari, et al. 1990. Diffusion-weighted MR imaging of anisotropic water diffusion in cat central nervous system. *Radiology* **176**: 439–445.

13   Stanisz, G.J., A. Szafer, G.A. Wright and R.M. Henkelman. 1997. An analytical model of restricted diffusion in bovine optic nerve. *Magn. Reson. Med.* **37**: 103–111.

14   Stejskal, E. 1965. Use of spin echoes in a pulsed magnetic-field gradient to study restricted diffusion and flow. *J. Chem. Physics* **43**: 3597–3603.

15   Stejskal, E.O. and J.E. Tanner. 1965. Spin diffusion measurement: spin echoes in the presence of a time-dependent field gradient. *J. Chem. Phys.* **42**: 288.

16   Tanner, J.E. 1979. Self diffusion of water in frog muscle. *Biophys. J.* **28**: 107–116.

17   Turner, R., D. LeBihan, J. Maier, R. Vavrek, L.K. Hedges, et al. 1990. Echo-planar imaging of intravoxel incoherent motions. *Radiology* **177**: 407–414.

18   van Gelderen, P., M.H. de Vleeschouwer, D. DesPres, J. Pekar, P.C.M. van Zijl, et al. 1994. Water diffusion and acute stroke. *Magn. Reson. Med.* **31**: 154–163.

19   van Gelderen, P., D. DesPres, P.C.M. van Zijl and C.T.W. Moonen. 1994. Evaluation of restricted diffusion in cylinders. Phosphocreatine in rat muscle. *J. Magn. Reson. B* **103**: 247–254.

20   van Zijl, P.C.M., D. Davis and C.T.W. Moonen. 1994. Diffusion Spectroscopy in Living Systems. NMR in Physiology and Biomedicine. R.J. Gillies (Ed.), San Diego, Academy Press: 185–198.

CHAPTER 3

1   Basser, P.J. and D.K. Jones. 2002. Diffusion-tensor MRI: theory, experimental design and data analysis — a technical review. *NMR Biomed.* **15**: 456–467.

2   Le Bihan, D., E. Breton, D. Lallemand, P. Grenier, E. Cabanis, et al. 1986. MR imaging of intravoxel incoherent motions: application to diffusion and perfusion in neurologic disorders. *Radiology* **161**: 401–407.

3   Stejskal, E. 1965. Use of spin echoes in a pulsed magnetic-field gradient to study restricted diffusion and flow. *J. Chem. Physics* **43**: 3597–3603.

4   Stejskal, E.O. and J.E. Tanner. 1965. Spin diffusion measurement: spin echoes in the presence of a time-dependent field gradient. *J. Chem. Phys.* **42**: 288.

CHAPTER 4

1   Basser, P.J., J. Mattiello and D. LeBihan. 1994. Estimation of the effective self-diffusion tensor from the NMR spin echo. *J. Magn. Reson. B* **103**: 247–254.

2   Basser, P.J. and D.K. Jones. 2002. Diffusion-tensor MRI: theory, experimental design and data analysis — a technical review. *NMR Biomed.* **15**: 456–467.

3   Beaulieu, C. 2002. The basis of anisotropic water diffusion in the nervous system — a technical review. *NMR Biomed.* **15**: 435–455.

4   Henkelman, R., G. Stanisz, J. Kim and M. Bronskill. 1994. Anisotropy of NMR properties of tissues. *Magn. Reson. Med.* **32**: 592–601.

5   Stanisz, G.J., A. Szafer, G.A. Wright and R.M. Henkelman. 1997. An analytical model of restricted diffusion in bovine optic nerve. *Magn. Reson. Med.* **37**: 103–111.

CHAPTER 5

1   Basser, P.J. and D.K. Jones. 2002. Diffusion-tensor MRI: theory, experimental design and data analysis — a technical review. *NMR Biomed.* **15**: 456–467.

2   Basser, P.J., J. Mattiello and D. Le Bihan. 1994. MR diffusion tensor spectroscopy and imaging. *Biophys. J.* **66**: 259–267.

3   Basser, P.J., J. Mattiello and D. Le Bihan. 1994. Estimation of the effective self-diffusion tensor from the NMR spin echo. *J. Magn. Reson. B* **103**: 247–254.

CHAPTER 6

1   Alexander, D.C. and G.J. Barker. 2005. Optimal imaging parameters for fiber-orientation estimation in diffusion MRI. *Neuroimage* **27**: 357–367.

2   Anderson, A.W. and J.C. Gore. 1994. Analysis and correction of motion artifacts in diffusion weighted imaging. *Magn. Reson. Med.* **32**: 379–387.

3   Bammer, R., M. Auer, S.L. Keeling, M. Augustin, L.A. Stables, et al. 2002. Diffusion tensor imaging using single-shot SENSE-EPI. *Magn. Reson. Med.* **48**: 128–136.

4   Bsatin, M 1999. Correction of eddy current-induced artifacts in diffusion tensor imaging using iterative cross-correlation. *Magn. Reson. Imaging* **17**: 1011–1024.

5   Butts, K., A. de Crespigny, J.M. Pauly and M. Moseley. 1996. Diffusion-weighted interleaved echo-planar imaging with a pair of orthogonal navigator echoes. *Magn. Reson. Med.* **35**: 763–770.

6   Clark, C.A., M. Hedehus and M.E. Moseley. 2002. In vivo mapping of the fast and slow diffusion tensors in human brain. *Magn. Reson. Med.* **47**: 623–628.

7   Clark, C.A. and D. Le Bihan. 2000. Water diffusion compartmentation and anisotropy at high b values in the human brain. *Magn. Reson. Med.* **44**: 852–859.

8   Hasan, K.M., D.L. Parker and A.L. Alexander. 2001. Comparison of gradient encoding schemes for diffusion-tensor MRI. *J. Magn. Reson. Imaging* **13**: 769–780.

9   Horsfield, M. 1999. Mapping eddy current induced fields for the correction of diffusion-weighted echo planar images. *Magn. Reson. Imaging* **17**: 1335–1345.

10   Jaermann, T., G. Crelier, K.P. Pruessmann, X. Golay, T. Netsch, et al. 2004. SENSE-DTI at 3T. *Magn. Reson. Med.* **51**: 230–236.

11   Jazzard, P., A.S. Barnett and C. Pierpaoli. 1998. Characterization of and correction for eddy current artifacts in echo planar diffusion imaging. *Magn. Reson. Med.* **39**: 801–812.

12   Jiang, H., X. Golay, P.C. van Zijl and S. Mori. 2002. Origin and minimization of residual motion-related artifacts in navigator-corrected segmented diffusion-weighted EPI of the human brain. *Magn. Reson. Med.* **47**: 818–822.

13   Jones, D.K. 2004. The effect of gradient sampling schemes on measures derived from diffusion tensor MRI: a Monte Carlo study. *Magn. Reson. Med.* **51**: 807–815.

14   Jones, D.K., M.A. Horsfield and A. Simmons. 1999. Optimal strategies for measuring diffusion in anisotropic systems by magnetic resonance imaging. *Magn. Reson. Med.* **42**: 515–525.

15   Kim, J., R. Kanaan, and G.D. Pearlson. 2002. Extended mutual information registration for simultaneously correcting motion effects and Eddy current distortion of single-shot echo-planar diffusion tensor imaging. *Human Brain Map.,* Sendai, Japan.

16   Maier, S.E., S. Vajapeyam, H. Mamata, C.F. Westin, F.A. Jolesz, et al. 2004. Biexponential diffusion tensor analysis of human brain diffusion data. *Magn. Reson. Med.* **51**: 321–330.

17   Mangin, J.F., C. Poupon, C. Clark, D. Le Bihan and I. Bloch. 2002. Distortion correction and robust tensor estimation for MR diffusion imaging. *Med. Image. Anal.* **6**: 191–198.

18   Mulkern, R.V., H. Gudbjartsson, C.F. Westin, H.P. Zengingonul, W. Gartner, et al. 1999. Multi-component apparent diffusion coefficients in human brain. *NMR Biomed* **12**: 51–62.

19   Norris, D.G. 2001. Implications of bulk motion for diffusion-weighted imaging experiments: effects, mechanisms, and solutions. *J. Magn. Reson. Imaging* **13**: 486–495.

20   Ordidge, R.J., J.A. Helpern, Z.X. Qing, R.A. Knight and V. Nagesh. 1994. Correction of motional artifacts in diffusion-weighted NMR images using navigator echoes. *Magn. Reson. Imaging* **12**: 455–460.

21   Pierpaoli, C. and P.J. Basser. 1996. Toward a quantitative assessment of diffusion anisotropy. *Magn. Reson. Med.* **36**: 893–906.

22   Pierpaoli, C., P. Jezzard, P.J. Basser, A. Barnett and G. Di Chiro. 1996. Diffusion tensor MR imaging of human brain. *Radiology* **201**: 637–648.

23   Pruessmann, K.P., M. Weiger, M.B. Scheidegger and P. Boesiger. 1999. SENSE: sensitivity encoding for fast MRI. *Magn. Reson. Med.* **42**: 952–962.

24   Skare, S., M. Hedehus, M.E. Moseley and T.Q. Li. 2000. Condition number as a measure of noise performance of diffusion tensor data acquisition schemes with MRI. *J. Magn. Reson.* **147**: 340–352.

25   Zhou, X. and Reynolds, H.G. 1997. Quantitative analysis of eddy current effects on diffusion-weighted epi. International Society for Magnetic Resonance in Medicine, Vancouver.

CHAPTER 7

1   Baratti, C., A. Barnett and C. Pierpaoli. 1999. Comparative MR imaging study of brain maturation in kittens with t1, t2, and the trace of the diffusion tensor. *Radiology* **210**: 133–142.

2   Beaulieu, C. 2002. The basis of anisotropic water diffusion in the nervous system — a technical review. *NMR Biomed.* **15**: 435–455.

3   Beaulieu, C. and P.S. Allen. 1994. Determinants of anisotropic water diffusion in nerves. *Magn. Reson. Med.* **31**: 394–400.

4   Conturo, T.E., R.C. McKinstry, E. Akbudak and B.H. Robinson. 1996. Encoding of anisotropic diffusion with tetrahedral gradients: A general mathematical diffusion formalism and experimental results. *Magn. Reson. Med.* **35**: 399–412.

5   Douek, P., R. Turner, J. Pekar, N. Patronas and D. Le Bihan. 1991. MR color mapping of myelin fiber orientation. *J. Comput. Assist. Tomogr.* **15**: 923–929.

6   Henkelman, R., G. Stanisz, J. Kim and M. Bronskill. 1994. Anisotropy of NMR properties of tissues. *Magn. Reson. Med.* **32**: 592–601.

7   Lazar, M., J.H. Lee and A.L. Alexander. 2005. Axial asymmetry of water diffusion in brain white matter. *Magn. Reson. Med.* **54**: 860–867.

8   Makris, N., A.J. Worth, A.G. Sorensen, G.M. Papadimitriou, T.G. Reese, et al. 1997. Morphometry of in vivo human white matter association pathways with diffusion weighted magnetic resonance imaging. *Ann. Neurol.* **42**: 951–962.

9   Mori, S., S. Wakana, L.M. Nagae-Poetscher and P.C. van Zijl. 2005. MRI atlas of human white matter. Amsterdam, The Netherlands, Elsevier.

10   Nakada, T. and H. Matsuzawa. 1995. Three-dimensional anisotropy contrast magnetic resonance imaging of the rat nervous system: MR axonography. *Neurosc. Res.* **22**: 389–398.

11   Neil, J., S. Shiran, R. McKinstry, G. Schefft, A. Snyder, et al. 1998. Normal brain in human newborns: apparent diffusion coefficient and diffusion anisotropy measured by using diffusion tensor MR imaging. *Radiology* **209**: 57–66.

12   Pajevic, S. and C. Pierpaoli. 1999. Color schemes to represent the orientation of anisotropic tissues from diffusion tensor data: application to white matter fiber tract mapping in the human brain. *Magn. Reson. Med.* **42**: 526–540.

13   Pierpaoli, C. and P.J. Basser. 1996. Toward a quantitative assessment of diffusion anisotropy. *Magn. Reson. Med.* **36**: 893–906.

14   Pierpaoli, C., P. Jezzard, P.J. Basser, A. Barnett and G. Di Chiro. 1996. Diffusion tensor MR imaging of human brain. *Radiology* **201**: 637–648.

15   Song, S.K., S.W. Sun, J.J. Roasbottom, C. Chang, J. Russell and A.H. Cross. 2002. Dysmyelination revealed through NMR as increased radial (but unchanged axial) diffusion of water. *Neuroimage* **17**: 1429–1436.

16   Stanisz, G.J., A. Szafer, G.A. Wright and R.M. Henkelman. 1997. An analytical model of restricted diffusion in bovine optic nerve. *Magn. Reson. Med.* **37**: 103–111.

17   Ulug, A.M., O. Bakht, R.N. Bryan and P.C.M. van Zijl 1996. Mapping of human brain fibers using diffusion tensor imaging. International Society for Magnetic Resonance in Medicine, New York.

18   Westin, C.F., S.E. Maier, H. Mamata, A. Nabavi, F.A. Jolesz, et al. 2002. Processing and visualization for diffusion tensor MRI. *Med. Image. Anal.* **6**: 93–108.

19   Zhang, J., L.J. Richards, P. Yarowsky, H. Huang, P.C. van Zijl, et al. 2003. Three-dimensional anatomical characterization of the developing mouse brain by diffusion tensor microimaging. *Neuroimage* **20**: 1639–1648.

20   Zhang, J., P.C. van Zijl and S. Mori. 2006. Image contrast using the secondary and tertiary eigenvectors in diffusion tensor imaging. *Magn. Reson. Med.* in press.

CHAPTER 8

1   Alexander, D.C., G.J. Barker and S.R. Arridge. 2002. Detection and modeling of non-Gaussian apparent diffusion coefficient profiles in human brain data. *Magn. Reson. Med.* **48**: 331–340.

2   Basser, P.J. and D.K. Jones. 2002. Diffusion-tensor MRI: theory, experimental design and data analysis — a technical review. *NMR Biomed.* **15**: 456–467.

3   Frank, L.R. 2001. Anisotropy in high angular resolution diffusion-weighted MRI. *Magn. Reson. Med.* **45**: 935–939.

4   Frank, L.R. 2002. Characterization of anisotropy in high angular resolution diffusion-weighted MRI. *Magn. Reson. Med.* **47**: 1083–1099.

5   Jansons, K.M. and D.C. Alexander. 2003. Persistent angular structure: new insights from diffusion magnetic resonance imaging data. *Inverse Problems* **19**: 1031–1046.

6   Lin, C.P., W.Y. Tseng, H.C. Cheng and J.H. Chen. 2001. Validation of diffusion tensor magnetic resonance axonal fiber imaging with registered manganese-enhanced optic tracts. *Neuroimage* **14**: 1035–1047.

7   Tournier, J.D., F. Calamante, D.G. Gadian and A. Connelly. 2004. Direct estimation of the fiber orientation density function from diffusion-weighted MRI data using spherical deconvolution. *Neuroimage* **23**: 1176–1185.

8   Tuch, D.S., T.G. Reese, M.R. Wiegell, N. Makris, J.W. Belliveau, et al. 2002. High angular resolution diffusion imaging reveals intravoxel white matter fiber heterogeneity. *Magn. Reson. Med.* **48**: 577–582.

9   Tuch, D.S., T.G. Reese, M.R. Wiegell and V.J. Wedeen. 2003. Diffusion MRI of complex neural architecture. *Neuron* **40**: 885–895.

10   Wedeen, V.J., P. Hagmann, W.Y. Tseng, T.G. Reese and R.M. Weisskoff. 2005. Mapping complex tissue architecture with diffusion spectrum magnetic resonance imaging. *Magn. Reson. Med.* **54**: 1377–1386.

11   Wiegell, M., H. Larsson and V. Wedeen. 2000. Fiber crossing in human brain depicted with diffusion tensor MR imaging. *Radiology* **217**: 897–903.

CHAPTER 9

1   Anderson, A.W. 2001. Theoretical analysis of the effects of noise on diffusion tensor imaging. *Magn. Reson. Med.* **46**: 1174–1188.

2   Basser, P.J., S. Pajevic, C. Pierpaoli, J. Duda and A. Aldroubi. 2000. In vitro fiber tractography using DT–MRI data. *Magn. Reson. Med.* **44**: 625–632.

3   Catani, M., R.J. Howard, S. Pajevic and D.K. Jones. 2002. Virtual in vivo interactive dissection of white matter fasciculi in the human brain. *Neuroimage* **17**: 77–94.

4   Conturo, T.E., N.F. Lori, T. Cull, S.E. Akbudak, A.Z. Snyder, et al. 1999. Tracking neuronal fiber pathways in the living human brain. *Proc. Natl. Acad. Sci. USA* **96**: 10422–10427.

5   Coulon, O., D.C. Alexander and S. Arridge. 2004. Diffusion tensor magnetic resonance image regularization. *Med. Image. Anal.* **8**: 47–67.

6   Huang, H., J. Zhang, P.C. van Zijl and S. Mori. 2004. Analysis of noise effects on DTI-based tractography using the brute-force and multi-ROI approach. *Magn. Reson. Med.* **52**: 559–565.

7   Jones, D.K. and C. Pierpaoli. 2005. Confidence mapping in diffusion tensor magnetic resonance imaging tractography using a bootstrap approach. *Magn. Reson. Med.* **53**: 1143–1149.

8   Jones, D.K., A. Simmons, S.C. Williams and M.A. Horsfield. 1999. Non-invasive assessment of axonal fiber connectivity in the human brain via diffusion tensor MRI. *Magn. Reson. Med.* **42**: 37–41.

9   Jones, D.K., A. Simmons, S.C.R. Williams and M.A. Horsefield. 1998. Non-invasive assessment of structural connectivity in white matter by diffusion tensor MRI. Sydney, Proceeding of International Society for Magnetic Resonance in Medicine.

10 Jones, D.K., A.R. Travis, G. Eden, C. Pierpaoli and P.J. Basser. 2005. PASTA: pointwise assessment of streamline tractography attributes. *Magn. Reson. Med.* **53**: 1462–1467.

11 Kinoshita, M., K. Yamada, N. Hashimoto, A. Kato, S. Izumoto, et al. 2005. Fiber-tracking does not accurately estimate size of fiber bundle in pathological condition: initial neurosurgical experience using neuronavigation and subcortical white matter stimulation. *Neuroimage* **25**: 424–429.

12 Lazar, M. and A.L. Alexander. 2002. White matter tractography using random vector (RAVE) pertubation. Honolulu, Proc. of 10th Annual ISMRM.

13 Lazar, M. and A.L. Alexander. 2003. An error analysis of white matter tractography methods: synthetic diffusion tensor field simulations. *Neuroimage* **20**: 1140–1153.

14 Lazar, M. and A.L. Alexander. 2005. Bootstrap white matter tractography (BOOT-TRAC). *Neuroimage* **24**: 524–532.

15 Lazar, M., D.M. Weinstein, J.S. Tsuruda, K.M. Hasan, K. Arfanakis, et al. 2003. White matter tractography using diffusion tensor deflection. *Hum. Brain Mapp.* **18**: 306–321.

16 Lori, N.F., J.S. Akbudak, T.S. Shimony, R.K. Snyder and T.E. Conturo. 2002. Diffusion tensor fiber tracking of brani connectivity: Reliability analysis and biological results. *NMR Biomed.* **15**: 494–515.

17 Mori, S., B.J. Crain, V.P. Chacko and P.C.M. van Zijl. 1999. Three dimensional tracking of axonal projections in the brain by magnetic resonance imaging. *Annal. Neurol.* **45**: 265–269.

18 Mori, S. and P.C. Van Zijl. 2002. Fiber tracking: principles and strategies - a technical review. *NMR Biomed.* **15**: 468–480.

19 Mori, S., S. Wakana, L.M. Nagae-Poetscher and P.C. van Zijl. 2005. *MRI atlas of human white matter.* Amsterdam, The Netherlands, Elsevier.

20 Parker, G.J. 2000. Tracing fiber tracts using fast marching. Denver, CO, International Society of Magnetic Resonance.

21 Parker, G.J., H.A. Haroon and C.A. Wheeler-Kingshott. 2003. A framework for a streamline-based probabilistic index of connectivity (PICo) using a structural interpretation of MRI diffusion measurements. *J. Magn. Reson. Imaging* **18**: 242–254.

22 Parker, G.J., K.E. Stephan, G.J. Barker, J.B. Rowe, D.G. MacManus, et al. 2002. Initial demonstration of in vivo tracing of axonal projections in the macaque brain and comparison with the human brain using diffusion tensor imaging and fast marching tractography. *Neuroimage* **15**: 797–809.

23   Parker, G.J., C.A. Wheeler-Kingshott and G.J. Barker. 2002. Estimating distributed anatomical connectivity using fast marching methods and diffusion tensor imaging. *IEEE Trans. Med. Imaging* **21**: 505–512.

24   Pierpaoli, C., A. Barnett, S. Pajevic, R.L. Chen, R. Penix, et al. 2001. Water diffusion changes in Wallerian degeneration and their dependence on white matter architecture. *Neuroimage* **13**: 1174–1185.

25   Poupon, C., C.A. Clark, V. Frouin, J. Regis, L. Bloch, et al. 2000. Regularization of diffusion-based direction maps for the tracking of brain white matter fascicules. *Neuroimage* **12**: 184–195.

26   Poupon, C., J. Mangin, C.A. Clark, V. Frouin, J. Regis, et al. 2001. Towards inference of human brain connectivity from MR diffusion tensor data. *Med. Image Anal.* **5**: 1–15.

27   Stieltjes, B., W.E. Kaufmann, P.C.M. van Zijl, K. Fredericksen, G.D. Pearlson, et al. 2001. Diffusion tensor imaging and axonal tracking in the human brainstem. *Neuroimage* **14**: 723–735.

28   Tournier, J.D., F. Calamante, D.G. Gadian and A. Connelly. 2003. Diffusion-weighted magnetic resonance imaging fibre tracking using a front evolution algorithm. *Neuroimage* **20**: 276–288.

29   Tournier, J.D., F. Calamante, M.D. King, D.G. Gadian and A. Connelly. 2002. Limitations and requirements of diffusion tensor fiber tracking: an assessment using simulations. *Magn. Reson. Med.* **47**: 701–708.

30   Tuch, D.S., J.W. Belliveau and V. Wedeen 2000. A path integral approach to white matter tractography. Proceeding of International Society of Magnetic Resonance in Medicine, Denver, CO.

31   Tuch, D.S., M.R. Wiegell, T.G. Reese, J.W. Belliveau, and V. Wedeen. 2001. Measuring cortico-cortical connectivity matrices with diffusion spectrum imaging. Proceeding of International Society of Magnetic Resonance in Medicine. UK, Glasgow.

32   Wakana, S., H. Jiang, L.M. Nagae-Poetscher, P.C. Van Zijl and S. Mori. 2004. Fiber Tract-based Atlas of Human White Matter Anatomy. *Radiology* **230**: 77–87.

33   Wedeen, V., T.G. Reese, D.S. Tuch, M.R. Weigel, J.G. Dou, et al. 2000. Mapping fiber orientation spectra in cerebral white matter with fourier-transform diffusion MRI. Proceeding of International Society of Magnetic Resonance in Medicine, Denvor, CO.

34   Werring, D.J., A.T. Toosy, C.A. Clark, G.J. Parker, G.J. Barker, et al. 2000. Diffusion tensor imaging can detect and quantify corticospinal tract degeneration after stroke. *J. Neurol. Neurosurg. Psychiatry* **69**: 269–272.

35   Xue, R., P.C.M. van Zijl, B.J. Crain, M. Solaiyappan and S. Mori. 1999. In vivo three-dimensional reconstruction of rat brain axonal projections by diffusion tensor imaging. *Magn. Reson. Med.* **42**: 1123–1127.

CHAPTER 10

1   Alexander, D.C., C. Pierpaoli, P.J. Basser and J.C. Gee. 2001. Spatial transformations of diffusion tensor magnetic resonance images. *IEEE Trans. Med. Imaging* **20**: 1131–1139.

2   Jones, D.K., L.D. Griffin, D.C. Alexander, M. Catani, M.A. Horsfield, et al. 2002. Spatial normalization and averaging of diffusion tensor MRI data sets. *Neuroimage* **17**: 592–617.

3   Lazar, M. and A.L. Alexander. 2005. Bootstrap white matter tractography (BOOT-TRAC). *Neuroimage* **24**: 524–532.

4   Pagani, E., M. Filippi, M.A. Rocca and M.A. Horsfield. 2005. A method for obtaining tract-specific diffusion tensor MRI measurements in the presence of disease: application to patients with clinically isolated syndromes suggestive of multiple sclerosis. *Neuroimage* **26**: 258–265.

5   Park, H.J., M. Kubicki, M.E. Shenton, A. Guimond, R.W. McCarley, et al. 2003. Spatial normalization of diffusion tensor MRI using multiple channels. *Neuroimage* **20**: 1995–2009.

6   Stieltjes, B., W.E. Kaufmann, P.C.M. van Zijl, K. Fredericksen, G.D. Pearlson, et al. 2001. Diffusion tensor imaging and axonal tracking in the human brainstem. *Neuroimage* **14**: 723–735.

7   Virta, A., A. Barnett and C. Pierpaoli. 1999. Visualizing and characterizing white matter fiber structure and architecture in the human pyramidal tract using diffusion tensor MRI. *Magn. Reson. Imaging* **17**: 1121–1133.

8   Xu, D., S. Mori, D. Shen, P.C van Zijl and C. Davatzikos. 2003. Spatial normalization of diffusion tensor fields. *Magn. Reson. Med.* **50**: 175–182.

9   Xue, R., P.C.M. van Zijl, B.J. Crain, M. Solaiyappan and S. Mori. 1999. In vivo three-dimensional reconstruction of rat brain axonal projections by diffusion tensor imaging. *Magn. Reson. Med.* **42**: 1123–1127.

CHAPTER 11

1   Field, A.S. and A.L. Alexander. 2004. Diffusion tensor imaging in cerebral tumor diagnosis and therapy. *Top Magn. Reson. Imaging* **15**: 315–324.

2   Hermoye, L., C. Saint-Martin, G. Cosnard, S.K. Lee, J. Kim, et al. 2005. Pediatric diffusion tensor imaging: Normal database and observation of the white matter maturation in early childhood. *Neuroimage.*

3   Holodny, A.I., D.M. Gor, R. Watts, P.H. Gutin and A.M. Ulug. 2005. Diffusion-tensor MR tractography of somatotopic organization of corticospinal tracts in the internal capsule: initial anatomic results in contradistinction to prior reports. *Radiology* **234**: 649–653.

4   Holodny, A.I. and M. Ollenschlager. 2002. Diffusion imaging in brain tumors. *Neuroimaging Clin. N. Am.* **12**: 107–124, x.

5   Horsfield, M.A. and D.K. Jones. 2002. Applications of diffusion-weighed and diffusion tensor MRI to white matter diseases. *NMR Biomed* **15**: 570–577.

6   Huppi, P., S. Maier, S. Peled, G. Zientara, P. Barnes, et al. 1998. Microstructural development of human newborn cerebral white matter assessed in vivo by diffusion tensor magnetic resonance imaging. *Pediatr. Res.* **44**: 584–590.

7   Jacobs, R.E., E.T. Ahrens, M.E. Dickinson and D. Laidlaw. 1999. Towards a microMRI atlas of mouse development. *Comput. Med. Imaging Graph.* **23**: 15–24.

8   Jacobs, R.E., E.T. Ahrens, T.J. Meade and S.E. Fraser. 1999. Looking deeper into vertebrate development. *Trends Cell Biol.* **9**: 73–76.

9   Lee, J.S., M.K. Han, S.H. Kim, O.K. Kwon and J.H. Kim. 2005. Fiber tracking by diffusion tensor imaging in corticospinal tract stroke: Topographical correlation with clinical symptoms. *Neuroimage* **26**: 771–776.

10   Lee, S.K., D.I. Kim, J. Kim, D.J. Kim, H.D. Kim, et al. 2005. Diffusion-tensor MR imaging and fiber tractography: a new method of describing aberrant fiber connections in developmental CNS anomalies. *Radiographics* **25**: 53–65; discussion 66–8.

11   Lee, S.K., S. Mori, D.J. Kim, S.Y. Kim and D.I. Kim. 2004. Diffusion tensor MR imaging visualizes the altered hemispheric fiber connection in callosal dysgenesis. *AJNR Am. J. Neuroradiol.* **25**: 25–28.

12   McGraw, P., L. Liang and J.M. Provenzale. 2002. Evaluation of normal age-related changes in anisotropy during infancy and childhood as shown by diffusion tensor imaging. *AJR Am. J. Roentgenol.* **179**: 1515–1522.

13   McKinstry, R.C., A. Mathur, J.H. Miller, A. Ozcan, A.Z. Snyder, et al. 2002. Radial organization of developing preterm human cerebral cortex revealed by non-invasive water diffusion anisotropy MRI. *Cereb. Cortex* **12**: 1237–1243.

14   Miller, S.P., D.B. Vigneron, R.G. Henry, M.A. Bohland, C. Ceppi-Cozzio, et al. 2002. Serial quantitative diffusion tensor MRI of the premature brain: development in newborns with and without injury. *J. Magn. Reson. Imaging* **16**: 621–632.

15   Mori, S., K. Fredericksen, P.C. van Zijl, B. Stieltjes, A.K. kraut, et al. 2002. Brain white matter anatomy of tumor patients using diffusion tensor imaging. *Annals of Neurol.* **51**: 377–380.

16   Mukherjee, P., M.M. Bahn, R.C. McKinstry, J.S. Shimony, T.S. Cull, et al. 2000. Differences between gray matter and white matter water diffusion in stroke: diffusion-tensor MR imaging in 12 patients. *Radiology* **215**: 211–220.

17   Mukherjee, P., J.H. Miller, J.S. Shimony, J.V. Philip, D. Nehra, et al. 2002. Diffusion-tensor MR imaging of gray and white matter development during normal human brain maturation. *AJNR Am. J. Neuroradiol.* **23**: 1445–1456.

18   Neil, J., J. Miller, P. Mukherjee and P.S. Huppi. 2002. Diffusion tensor imaging of normal and injured developing human brain — a technical review. *NMR Biomed.* **15**: 543–552.

19   Neil, J., S. Shiran, R. McKinstry, G. Schefft, A. Snyder, et al. 1998. Normal brain in human newborns: apparent diffusion coefficient and diffusion anisotropy measured by using diffusion tensor MR imaging. *Radiology* **209**: 57–66.

20   Partridge, S.C., P. Mukherjee, R.G. Henry, S.P. Miller, J.I. Berman, et al. 2004. Diffusion tensor imaging: serial quantitation of white matter tract maturity in premature newborns. *Neuroimage* **22**: 1302–1314.

21   Schneider, J.F., K.A. Il'yasov, J. Hennig and E. Martin. 2004. Fast quantitative diffusion-tensor imaging of cerebral white matter from the neonatal period to adolescence. *Neuroradiology* **46**: 258–266.

22   Song, S.K., S.W. Sun, M.J. Ramsbottom, C. Chang, J. Russell, et al. 2002. Dysmyelination revealed through MRI as increased (radial but unchanged axial) diffusion of water. *Neuroimage* **17**: 1429–1436.

23 Stieltjes, B., W.E. Kaufmann, P.C.M. van Zijl, K. Fredericksen, G.D. Pearlson, et al. 2001. Diffusion tensor imaging and axonal tracking in the human brainstem. *Neuroimage* **14**: 723–735.

24 Sun, S.W., J.J. Neil, H.F. Liang, Y.Y. He, R.E. Schmidt, et al. 2005. Formalin fixation alters water diffusion coefficient magnitude but not anisotropy in infarcted brain. *Magn. Reson. Med.* **53**: 1447–1451.

Subject Index

Printed and bound by CPI Group (UK) Ltd, Croydon, CR0 4YY

03/10/2024

01040728-0003